WINDFALL

WINDFALL

THE BOOMING BUSINESS
OF GLOBAL WARMING

McKENZIE FUNK

THE PENGUIN PRESS

New York

2014

THE PENGUIN PRESS
Published by the Penguin Group
Penguin Group (USA) LLC
375 Hudson Street
New York, New York 10014

USA · Canada · UK · Ireland · Australia
New Zealand · India · South Africa · China

penguin.com
A Penguin Random House Company

First published by The Penguin Press, a member of Penguin Group (USA) LLC, 2014

Photographs by the author

ISBN 978-1-59420-401-2

Printed in the United States of America
1   3   5   7   9   10   8   6   4   2

Designed by Nicole LaRoche

For Jenny and Wilson.
(Mostly, she says, for him.)

CONTENTS

PART THREE

THE DELUGE

WINDFALL

INTRODUCTION

The contract had called for either a boa or an anaconda, whichever would best handle the crowds, and in the end the bankers got the latter: a green anaconda, six feet long and eighty-five pounds, which hung from the neck of a long-haired snake handler who lurked amid the exotic plants, next to the fake waterfall and the model dressed in "Amazonian" garb. Nearby were two scarlet macaws in wire cages, a Brazilian dance troupe, and a hut offering free organic smoothies. At the base of an eighteen-foot waterfall were giant koi, swimming in a pond: forty-five hundred gallons of warm, filtered water that would soon be dumped into the East River. The jungle was in a tent that was on the promenade at the South Street Seaport, in lower Manhattan. Thirty by sixty feet, suffused with a light mist, and heated to eighty degrees, the tent had white sides and a clear roof through which visitors could just make out the skyscrapers of Wall Street. It was cold outside, a typical thirty-nine-degree February day in early-twenty-first-century New York, so those beckoned inside by the street team—two models walking the streets to entice passersby to the event—had to quickly shed their jackets and scarves, so stark was the difference in temperature. Which was, of course, the point.

The stunt was a coming-out party, the most expensive stop on Deutsche Bank's eighty-event "The Investment Climate Is Changing" road show held across the United States. In scale and imagination, it was rivaled only by the ski village and ninety-foot snowboard slope the bank had constructed a few weeks earlier along Rodeo Drive in Beverly Hills: chalets decorated with deer-antler chandeliers and wooden snowshoes, Deutsche Bank–branded ice sculptures, models dressed as snow bunnies, bottled water from Iceland, faux snow blown down from the roof of the Versace store, thirty tons of more realistic snow created by a wood chipper and a freezer truck full of ice blocks, and two pro snowboarders who would later complain that nobody had built them a proper jump. Together, the Manhattan and Beverly Hills events cost $1.5 million, but they were carbon neutral, the bankers boasted, their greenhouse emissions offset by investments in a biogas project in India. At South Street Seaport, every attendee was given a certificate from the Carbon Credit Company as proof. The jungle party, which lasted three hours, produced 152 tons of greenhouse gases, which the average Indian would need three lifetimes to match.

Before a DJ set by the Brazilian Girls—a group with no actual Brazilians and only one girl—the bankers held a press conference. It was early 2008, and as the world was still reeling from a record melt in the Arctic and a scary film by Al Gore and a bleak report by the Intergovernmental Panel on Climate Change (IPCC), half a dozen major investment houses had launched global-warming-themed mutual funds. Deutsche Bank's was the $2.9 billion DWS Climate Change Fund. The jungle event was meant to promote it. "Without taking a position on climate change," a press release had explained, the "DWS Climate Change Fund is on the cutting edge of climate change investing." The event's objective was "not simply to show that climate change is happening," said the executive Axel Schwarzer, "but that it creates related climate change investment opportunities." Another release went further. "The debate around climate change is shifting away from cost and risk," it said, "toward the question of how to capitalize on exciting opportunities." Nothing as big and universal as climate change could be all bad. An ecological catastrophe was not necessarily a financial catastrophe for everyone.

Deutsche Bank's chief climate strategist, Mark Fulton, worked in midtown in a building on Park Avenue, and I visited him there after the road show was done, clearing security and then riding a silent elevator to the twenty-seventh floor. His was a corner office, but it was small and cluttered with papers, and Fulton, an Oxford-educated Australian, looked as much scientist as capitalist. His desire to fight climate change was genuine. He told me he'd read the Club of Rome's *Limits to Growth*—a neo-Malthusian take on the planet's carrying capacity—as a schoolboy in the 1970s. "It made quite an impact," he said. "They were talking about everything running out: 'What are we going to do? We have to change the way we live!'" Instead of working for Greenpeace, which he'd considered after graduation, he became a stockbroker, then an analyst, and he'd eventually helped Deutsche Bank identify global warming as a "megatrend" that could generate profits for decades. "It's always helped me, climate change, in my career," he joked.

While the DWS fund invested most heavily in the technology to build a greener world—in wind power and solar power, in smart grids and smarter electrical meters—it had bought other stocks, too: companies that fit the portfolio not because they could help fight climate change but because the warmer the world, the less habitable it became, the bigger the windfall. They were a tacit recognition that we were already failing to stop climate change. There was the planet's largest water company, Veolia, which manages pipes and builds desalination plants in seventy-four countries on five continents. Monsanto and Syngenta, ag-biotech giants that were tweaking genes to develop drought-resistant crops. And Viterra, a fast-growing agribusiness in temperate Canada. The fund also had shares of Duoyuan Global Water, one of the biggest water-treatment companies in desiccating China, and two fertilizer multinationals, Yara and Agrium. When I asked Fulton how the bank planned to capitalize on rising sea levels, he mentioned a small play in a Dutch dredging company, Royal Boskalis, which had just rebuilt an island in the Maldives inundated by the 2004 tsunami. "Where are you going to get seawall expertise but from the Dutch?" he asked.

Other climate investors told a similar story. They bought clean tech, green tech, the building blocks of the new, low-carbon economy—but they were also starting to hedge. In London, the Schroder Global Climate Change Fund was investing in Russian farmland—cheap, fertile soil suddenly made dear by milder winters and drought-fueled global food crises—and its manager was taking the logic a step further, buying stock in supermarket chains such as Carrefour and Tesco. "If climate change will be a negative for crop yields," he told me, "then people will just have to spend more on food. Retailers are a clear beneficiary." Across town, another fund manager explained why he was bullish on the reinsurers Munich Re and Swiss Re. "As natural disasters start to be more common," he said, "as climate change starts to cause flooding and droughts, insurance companies—reinsurers in particular—should get pricing power." Because it allows insurers to jack up rates, "hurricane season is actually quite a positive thing." A partner at a storied Wall Street investment bank showed me photographs of Ukrainian farmland and said his firm had tried to buy up "vast tracts" of it. Soviet-era collective farms had reverted to "pseudo-subsistence agriculture," he said. "You could come to these guys and get thousands of hectares for a few bottles of vodka and, like, two months of grain. You could literally give them vodka and grain."

In the run-up to successive climate conferences in Copenhagen, Cancún, Durban, and Doha, as everyone else was fretting about polar bears and electric cars, some fund managers worried I would misunderstand them—that I would mistake them for starry-eyed activists, that I would mistake theirs for just another green or socially responsible fund. "A lot of people think, 'How do you invest in climate change?' and essentially come up with one or two or maybe three areas, like alternative energy," Sophie Horsfall, a manager of Britain's F&C Global Climate Opportunities Fund, told me. "For us, well, there is an awful lot more to it. We have to separate out the ethical values. We have to move away from the environmental issues. We have to take a step back." I must have looked puzzled. "We have to think about the reality of climate change," she continued. "It is quite difficult, isn't it?"

FOR DECADES WE have all known, at some level, about global warming. As a point of scientific inquiry it is decades old, first identified in the 1800s by John Tyndall and Svante Arrhenius, but as a source of popular anxiety and conversation it dates to the first sophisticated computer models of the early 1970s and the first World Climate Conference in 1979 and landmark congressional testimony by the NASA atmospheric physicist James Hansen in 1988. It has been around long enough to become a cliché—I thank it for the heat wave I'm experiencing in Seattle as I write this—and long enough to have birthed a newer cliché: the idea that we have so changed the planet with our engineering and our emissions that we now live in the Anthropocene, a new geologic epoch of man's own creation. Long enough, certainly, for something to have been done about it. In the new millennium, which has brought us Al Gore's *Inconvenient Truth*, Lord Nicholas Stern's seven-hundred-page *Economics of Climate Change*, and a string of failed climate legislation and UN conferences, the warnings have been ever louder and more sustained. The atmospheric concentration of carbon dioxide, our principal contribution to the climate and the principal driver of warming, has only been rising. It is now 40 percent higher than preindustrial levels, higher than it has been anytime in the last 800,000 years. In New York's Madison Square Garden, a seventy-foot doomsday clock, recently unveiled by Deutsche Bank, is tracking greenhouse-gas levels in real time: 2 billion metric tons added each month, or 800 a second, for a total of 3.7 trillion tons and counting. The ticker has thirteen red digits, but when you stare at it from Seventh Avenue, the last three are a blur. They're spinning too quickly to see.

This book is about how we're preparing for the world we seem hell-bent on creating. It's about climate change, but not about the science of it, nor the politics, nor directly about how we can or why we should stop it. Instead, it's about bets being placed on a simple, cynical premise: that we won't stop it anytime soon. It's about people, and mostly it's about people like me: northerners from the developed world—historically the emitter countries, as we're called—who occupy the high, dry ground, whether real or metaphorical.

I'm interested in climate change as a driver of human behavior—as a case study, the ultimate case study, in how we confront crisis. Warming will reshape the planet, and in broad strokes we already know how: Hot places will get hotter. Wet places will get wetter. Ice will simply melt. Poor, mostly tropical countries, those least responsible for the consumption that fuels the factories that produce the emissions that cause the warming, will be hit hardest, but wealthier, higher-latitude regions—Europe, Canada, the United States—are not entirely immune. The change is so vast, so universal, that it seems to test the limits of human reason. So it should not be surprising that the ideologies that led us here, those that have guided the postindustrial age—techno-lust and hyper-individualism, conflation of growth with progress, unflagging faith in unfettered markets—are the same ones many now rely on as we try to find a way out. Nowhere is humankind's mix of vision and tunnel vision more apparent than in how we're planning for a warmed world.

The idea that people are irrational has lately been in vogue. We can thank the global financial crisis for that. Behavioral economists have reminded us that the market, far from being a collection of fully logical individuals, is hostage to Keynesian "animal spirits," the emotions, prejudices, impulses, and shortcuts that are part of nearly every human decision and every financial bubble—and part, no doubt, of our apathy about reducing carbon emissions. In the United States, nearly 98 percent of the federal climate-research budget goes to the hard sciences, which have produced mounds of evidence for global warming—enough to make a believer of anyone who gives it an honest look—and produced increasingly refined computer models predicting an increasingly dire future. One recent prediction, from MIT, is of a median warming of 5.2 degrees Celsius by 2100 if we don't curtail emissions—a temperature spike that campaigners believe could entirely melt the polar ice cap in summertime, turn parts of Central America and the southern United States into a dust bowl, and wipe island nations off the map. The remaining 2 percent of the federal research budget goes to social scientists, such as those with Columbia University's Center for Research on Environmental Decisions, who probe what may now be the most important question: If we know the risks, why aren't we doing anything? The center's director, Elke Weber, suggests that at both levels where humans make their decisions—emotional and analytical—there are roadblocks. The emotional block: What we don't see doesn't scare us. "The time-delayed, abstract, and often statistical nature of the risks of global warming does not evoke strong visceral reactions," Weber writes. At the analytical level, there is, along with the tension between individual and systemic risk—an apparent tragedy of the commons—something economists call hyperbolic discounting. It goes like this: Offer to give someone either $5 today or $10 next year, and he'll probably take the $5.

Among many activists, politicians, and scientists, the assumption is that climate change now suffers mainly from a PR problem: If the proper nudges can be found or the reality of it finally made visceral, the public will take action. Unspoken and scarcely examined is a second, much bigger assumption: that "taking action" means trying to cut carbon emissions. That taking action will take a certain shape: Green roofs. Carbon caps. Green cars. Solar panels. Footpaths. Forests. Fluorescent bulbs. Bicycles. Insulation. Algae. Inflated tires. Showers. Clotheslines. Recycling. Locavorism. Light-rail. Wind farms. Vegetarianism. Heat pumps. Telecommuting. Smaller homes. Smaller families. Smaller lives. We hope our collective fear of global warming will push us inevitably toward collective behavior. But what if the world as we know it goes on even as the Earth as we know it begins to disappear? There's another possible response to melting ice caps and rising sea levels, to the reality of climate change—a response that is tribal, primal, profit-driven, short-term, and not at all idealistic. Every man for himself. Every business for itself. Every city for itself. Every country for itself. There's the possibility that we take the $5.

SPEND AN AFTERNOON in the right part of the Arctic, perhaps in the company of a Russian or an Icelander or an oil executive, listen to the plans being hatched, and you can experience anew the carnival atmosphere of Deutsche Bank's jungle tent. The Arctic was where I did the first reporting for this book, and it was where I caught my first whiff of giddiness about climate change, of opportunism amid environmental crisis. There was oil under the ice. There were new shipping lanes emerging over the pole. There were strawberries sprouting in Greenland. The high north was the first place where warming had become not an invisible menace but a daily reality, thus the first place where I could actually witness people's reactions to it. I began traveling the rest of the globe with the same intent—to document present-day preparations for a warmer world, to observe what was happening rather than theorize about what could happen.

Global warming's physical impacts, the impetus for the plans and projects I investigated, can be separated into three broad categories: melt, drought, and deluge. Accordingly, this is a book in three parts. Part One, "The Melt," is set against the liquefaction of the world's ice sheets and glaciers, a process that is only accelerating: In recorded history, the Northwest Passage and the Northeast Passage have never, until today, become ice-free and thus open to commercial shipping, and the Arctic ice cap has never been smaller than it was in the summers of 2007, 2008, 2009, 2010, 2011, and especially 2012, when 4.57 million square miles, an area larger than the United States, melted away. Part Two, "The Drought," discusses the large-scale reordering of our planet's hydrology such that rain falls at different times, in different places, and deserts appear where there were none. In some places, drought is a result of melt; mountain snowfields and glaciers are the planet's best natural water reservoirs, and they are dramatically receding. That the drought is already beginning is evidenced not by specific events but by a pattern of them: wildfires in Colorado, water woes in northern China, desertification in Spain, food riots in Senegal, and the fact that to describe the recent state of Australia's breadbasket, the Murray-Darling basin, the term "drought" was discarded in favor of the more permanent-sounding "dryness." Part Three, "The Deluge," addresses what is generally our most distant concern, decades if not centuries out—the rising seas, surging rivers, and superstorms that will threaten island nations and coastal cities. But it is hastened as parched cities drain their aquifers and begin to sink, accelerated as Greenland's ice cap melts into the sea. And after Hurricane Sandy and Typhoon Bopha and failure after failure to cut global carbon emissions, it is not entirely distant.

To explore these changes in order—from melt to drought to deluge— as I broadly did, with some exceptions, in my travels around the globe, is to go from opportunism to wagon circling to open desperation. The expansionist exuberance of the Arctic petroleum rush, which has men running around like Elizabethan invaders, claiming virgin territory, fades into the grim free marketeering of a Malthusian world without enough water, then into the bunker mentality of sea-level rise and hurricanes, which could be what finally makes climate change personal for many Americans—and against which long-shot technology is viewed as our only escape hatch. There is no single response to the effects of global warming, even if we do seem to fall back on a finite set, but as I traveled, I found a consistent theme: I met hundreds of people who thought climate change would make them rich. In the six years I spent reporting this book, visiting twenty-four countries and more than a dozen U.S. states (and flying so often that I caused far, far more than my share of carbon emissions), I met profiteers, engineers, warlords, mercenaries, vigilantes, politicians, spies, entrepreneurs, and thieves—people seeking to come out ahead in a new, warmed world. They were universally kind and hospitable to me, and nearly all, driven by ideology, fear, or hard-nosed realism—or all three—thought they were doing the necessary thing. In six years, I never met a bad person.

When you're on the high ground—wealthy enough, northerly enough, far enough above the sea—global warming is not yet the existential threat that it is for an Egyptian or a Marshall or Staten Islander. It's a shorter ski season, a more expensive loaf of bread, a new business opportunity. We can afford the desalination plants; we can afford the seawalls. Many of the world's existing imbalances seem only magnified by climate change, and they may be magnified all the more by how we respond to it. The technical term for trying to prepare for an altered planet is "adaptation." (To try to cut emissions is known as mitigation.) One of the few tangible results of the 2009 and 2010 climate conferences in Copenhagen and Cancún was a pledge by emitter countries to help poorer countries adapt. But new climate funding is already falling short of the pledge: so far, $2 billion to save the rest of the world, which is at least $8 billion less than it could cost to build a proposed storm-surge barrier to protect New York City from the next Sandy.

It would be a mistake to suggest that every plan and project described in this book was born solely, or even principally, in response to climate change. Arctic oil is attractive for many reasons, not least because there's less and less oil everywhere else, and what remains is often in hostile countries (Iran, Venezuela, Sudan) or recent conflict zones (Iraq, Nigeria, Libya). Water markets have boomed in Australia and California thanks in large part to the historic oddities of their water laws and the decision, whether foolish or brave, to turn emptiness into farmland, deserts into paradise. African refugees crowding southern Europe's detention camps have often fled more immediate threats than the expanding Sahara. Genetic engineers racing to build supernaturally perfect corn see climate change as just one more excuse for their efforts. Weather modifiers have tried to make rain and tame hurricanes for a generation. The twenty-one-hundred-mile fence that India is building around Bangladesh is not all about sea-level rise, not hardly: India also doesn't much like Bangladesh, and its emigrants have long been a source of irritation. It is as difficult to attribute human action to a single climatic cause as it is to attribute today's weather report—or one bad wheat harvest—to long-term climatic shifts. But global warming is the thread that ties these stories together, and it's a window into our collective state of mind. I've tried to keep rooted in the present, so if there's a glimpse of the future in these pages, it's only because we're the ones making it. To the increasingly urgent question "What *are* we doing about climate change?" this book is meant to be an answer.

PART ONE

THE MELT

It is natural to expect that opinions were very varied when the news spread that the Arctic region was going to be sold at auction for the benefit of the highest and final bidder. . . .

To use the Arctic region? Why, such an idea could "only be found in the brain of a fool," was the general verdict.

Nothing, however, was more serious than this project.

—*Jules Verne*, The Purchase of the North Pole, *1889*

ONE

COLD RUSH

CANADA DEFENDS THE NORTHWEST PASSAGE

On the first full day of the sovereignty operation, the captain slowed the frigate and we took out the machine guns and sprayed the Northwest Passage with bullets. It felt pretty good. It was foggy out, and the unpolluted water boiled as we polluted it with lead. There was no life we could see and few waves. The wind was cold, the Arctic Ocean a drab green. There wasn't any ice. But if there had been ice, we would have shot it, too.

The guns were C7s—American M16s but rechristened, like many Canadian weapons, with a patriotic C—and most of the shooters were camo-clad teenagers from Quebec's celebrated 22ᵉ Regiment, who are known as the Vandoos, from *vingt-deux* (twenty-two). The Vandoos lined up three in a row on the back deck, each of them held in place by a sturdy navy man, and fired away. They went from semiautomatic to fully automatic and shot more. They switched to pistols and then shotguns and shot until the deck was littered with shells. When they finished, they kicked the shells into the sea. There were journalists on board, and the Arctic was melting, and the Canadians—who now had a new, northern coastline to develop and defend—were trying their hardest to be fierce. The world had to understand that they were ready to fight for whatever riches the retreating ice revealed.

The frigate was named the *Montreal*. It was the length of two city blocks and painted warship gray, packed with two dozen torpedoes and nearly 250 people. There were sailors, Vandoos, and Mounties. There were Canadian wire-service reporters and photographers from at least two in-flight magazines. There were Inuit dignitaries and observers from Nunavut Tunngavik Incorporated, the pseudo-governmental Inuit corporation that had negotiated the 1999 creation of its people's own 800,000-square-mile territory, Nunavut. Our cruise speed was 15.5 knots. Our fuel stores were at 125 percent. With diesel taking the place of water in the auxiliary tanks, our showers were capped at two minutes. We were steaming north, farther north than the Royal Canadian Navy had gone in decades.

The Arctic held two main prizes: petroleum and new shipping lanes. An estimated 22 percent of the world's untapped deposits—ninety billion barrels of oil and 1,670 trillion cubic feet of natural gas, according to the U.S. Geological Survey—is thought to be hiding in the high north, some of it in territory that does not yet belong to any nation. The less ice there is, the more petroleum there is within reach, and the more pressure there is to stake a claim. Likewise, the less ice there is, the more the storied Northwest Passage—a long-sought, long-frozen-over shortcut between the Atlantic and the Pacific—becomes a viable alternative to the Panama Canal, saving potentially shippers leaving Newark or Baltimore for Shanghai or Busan some four thousand miles and hundreds of thousands of dollars in transit fees and fuel costs.

Canada owns the land on both sides of the Northwest Passage, but much of the world, particularly its customary ally the United States, does not agree that it owns the waterway itself. Canadians were tired of being pushed around by their more populous neighbor—of being "condemned to always play 'Robin' to the U.S. 'Batman,'" as American diplomats would put it in a 2008 cable released by WikiLeaks. At stake up here was national pride, not just money or national security. To kick off this show of force, called Operation Lancaster, conservative prime minister Stephen Harper himself had made the long journey to Iqaluit, the former U.S. military base that is now the capital of Nunavut. He had arrived bearing promises of new heavy icebreakers, a new Arctic warfare and training center, a new deepwater port, and a new network of undersea sensors and aerial drones. Now, as his Vandoos and Mounties moved north, he was putting boots on the ice.

There had been sovereignty operations before, including Nunalivut (Inuktitut for "the land is ours") in 2006 and the previous year's Exercise Frozen Beaver, when Canadian troops helicoptered to Hans Island—a bean-shaped, half-square-mile rock near Greenland claimed by both Denmark and Canada—and planted a supposedly windproof steel flag and flagpole that the wind toppled almost immediately. But Lancaster was the largest such operation to date, the first to take advantage of retreating sea ice, and it was occurring on the hundredth anniversary of the Northwest Passage's first crossing (which was by a Norwegian, though no one dwelled on that). Its stated goal was to "project a credible size military force over a broad area of the Eastern Arctic." It would last twelve days in all. The *Montreal* would lead a flotilla of two navy warships and two coast guard icebreakers into Lancaster Sound, the eastern entrance of the passage, and patrol back and forth as the skies buzzed with Aurora surveillance planes and Griffon helicopters. Meanwhile, the Vandoos—accompanied by Inuit reservists, there to ensure that no one was eaten by polar bears—would take the smaller ships to shore and set up observation posts on both sides of the sound. To the north, on rocky Devon Island, would be Observation Post 1. To the south, on glaciated Bylot Island and the adjacent Borden Peninsula, would be Observation Posts 2 and 3. The troops would hold the high ground for most of a week, scanning the Northwest Passage for invaders.

This would all be preceded by a display of Canadian resolve: a mock interdiction. After watching the machine guns fire and the Maple Leaf flag flutter, I strolled up to the bridge and stood next to the *Montreal*'s commanding officer. He and his crew had donned green helmets and green flak jackets. The radio crackled, and a Canadian approximation of the voice of a California surfer filled the bridge. It was the supposed captain of the *Killer Bee,* which in actuality was the *Goose Bay,* a 150-foot Canadian coastal-defense vessel that the war gamers had decided would be a rogue "American" merchant ship starting an unauthorized transit of the Northwest Passage.

The *Killer Bee* was four miles away in the fog, sailing a course that would intersect with ours in an estimated fourteen minutes and forty-two seconds. It would not say where it was going. It would not say what was in its hold. "Merchant vessel *Killer Bee,* what is your cargo?" our radioman asked. "This is Warship 336. Again, what is your cargo?" The *Killer Bee*'s answers were brief, rude, believably American in their tone save for the occasional slipup: "We're aboot forty miles off the coast, which constitutes international waters. Are you sure you have the authority to be questioning me out here? Can you just tell me again why I'm being asked these questions? You guys are the almighty Canadian government, so I'm sure you can access this sort of information some-where else."

The *Montreal* passed a message to the colonel running Operation Lancaster, asking for clearance to send over a boarding party and, if necessary, to initiate "disabling fire." The sailors on the bridge peered into the mist off our port side. We informed the *Killer Bee* that we would be boarding it, and its captain replied that he wouldn't be "too down with that." The engine churned. We began to close the gap: seven hundred yards, six hundred yards, five hundred yards. The ship appeared, and we aimed our .50-caliber machine gun at it. "Bullying your way around the ocean is not a way to foster cooperation between our two countries," the voice told us. We commanded the *Killer Bee* to remove all personnel from its top decks, and our gunners directed a barrage of tracer fire a thousand yards off its bow. The smell of gunpowder wafted through the bridge. The next barrage was five hundred yards off the bow. Finally, our 57-millimeter cannon swiveled toward the *Killer Bee*. There were five loud booms in quick succession, five puffs of smoke, and then, seconds later, a sixth round. The ocean in front of the *Killer Bee* erupted. Its captain relented. "I thought Canada was a nation of peacekeepers," whined the faux American.

For the next five hundred miles, we saw only water and fog and an occasional glimpse of the chutes and pinnacles of Baffin Island's peaks. It wasn't until 10:00 a.m. on the operation's fourth day that a much-awaited announcement came over the loudspeaker: icebergs ahead. We rushed to the port-side deck where the officers normally gathered to smoke. We were at seventy-two degrees north, and there were three of them: two- and three-hundred-foot giants that towered over the frigate. The icebergs' walls were riven by small waterfalls, and chunks of ice were falling off into the sea. The bergs were drifting south toward the Atlantic, bound for warmer waters, where they would soon melt into nothing. The Vandoos leaned over the railing and snapped photographs.

IT WAS THE SUMMER OF 2006, and drought-crazed camels would soon rampage through a village in Australia, a manatee would swim past Chelsea Piers in New York City's Hudson River, and the Netherlands would announce that its famous Elfstedentocht ice-skating race might have to be postponed forever. Armadillos were reaching northeast Arkansas. Wolves ate dogs in Alaska. Fire consumed fifty million acres of Siberia. Greenland lost a hundred gigatons of ice. The Inuit got air-conditioning units. The polar bear lurched toward the endangered-species list. India's Ghoramara Island was mostly lost to the Bay of Bengal, Papua New Guinea's Malasiga village was mostly lost to the Solomon Sea, and Alaska's Shishmaref village decided to evacuate before being lost to the Chukchi Sea. Canadian scientists reported that the forty-square-mile Ayles Ice Shelf had broken off Ellesmere Island and formed a rapidly melting island of its own. A European satellite showed a temporary crack in the ice pack leading from northern Russia all the way to the North Pole. The National Oceanic and Atmospheric Administration would declare that winter the warmest since it began keeping records, which was in 1880. The Intergovernmental Panel on Climate Change would announce that eleven of the previous twelve years were the warmest in human history.

In retrospect, this was the moment that we began to believe in global warming—not in the abstract science of it, which most could already passively accept, but in the fact that there were money and power to be won and lost. Skeptics would continue loudly doubting the overwhelming scientific consensus, but they were a smoke screen. For those who considered climate change's strategic rather than ideological impacts—militaries, corporations, the rare politician—it had become time to grapple with the consequences. There would be winners. There would be losers. The process of determining who was who was getting under way.

Great Britain had recently asked its chief economist, Sir Nicholas Stern, to conduct a review of global warming's likely effects on world markets. His findings were dire: The cost of unchecked greenhouse-gas emissions would be the equivalent of losing 5 percent or more of global GDP a year, every year, forever. In tropical Africa and South America, crop yields would drop dramatically. In South and East Asia, hundreds of millions of people and trillions of dollars of assets would be threatened by rising seas. "What makes wars start?" Britain's foreign minister, Margaret Beckett, asked the UN Security Council in 2007. "Fights over water. Changing patterns of rainfall. Fights over food production, land use." According to Lord Stern, the world was on the brink of an upheaval on the scale of the two world wars and the Great Depression.

But the future did not seem universally dark. At the margins of the crisis, some were already seeing opportunity, especially in the wealthy nations that were causing climate change in the first place. At least in the near term in most of Europe, Russia, Canada, and America, rain will still fall, growing seasons will extend, and some agriculture could expand, bolstered by our emissions. Carbon dioxide is a key building block for plant growth. All else being equal—though in few cases will all else be equal—the higher the atmospheric concentration, the higher the yields.

Farther north, in the Arctic, the ice albedo feedback effect—the fact that sea ice, which reflects 85 to 90 percent of solar radiation, melts to become seawater, which absorbs all but 10 percent of radiation—would help keep temperatures climbing at twice the global rate. Northern economies seemed poised to grow at least as rapidly. Canada's farmers already had two extra growing days a year, and studies said its Athabasca tar sands might someday be accessible from the north, via the Mackenzie River. Under Stephen Harper, a country many Americans considered well-meaning to the point of naïveté was becoming one of the villains of international climate conferences. Canada was a party to the Kyoto Protocol, a weak 1997 treaty that mostly excluded big emitters like China and the United States but nonetheless remains the world's first and only binding international agreement on greenhouse gases. Yet Canada would be overshooting its Kyoto targets by 30 percent by the time it abandoned the treaty in 2012—just before another northern economy, Russia, also made its exit. One could blame Canada's climate about-face on its reliance on carbon-intensive tar sands. But it is also unclear that climate change is all that bad for Canada.

The $49 million grossed by Al Gore's *Inconvenient Truth* might have been global warming's first true financial success story, but as the *Montreal* entered the Northwest Passage, the new mentality was taking hold. Reports by Citigroup, UBS, and Lehman Brothers advised investors on how to wring a buck out of global tailspin. Citigroup's report *Climatic Consequences: Investment Implications of a Changing Climate*, released in January 2007, was particularly helpful. It highlighted investment opportunities at seventy-four companies in twenty-one industries in eighteen countries, including Aguas de Barcelona (drought-afflicted

Spain's "leader in water supply"), Monsanto (drought-resistant crops), and John Deere (more tractors needed in America as drought wiped out Australia's wheat exports). It showed a graph of the six top natural-gas-producing countries in the world. Four of them—Russia, the United States, Canada, and Norway—were Arctic nations.

MY BUNK MATE on the *Montreal* was a man I'll call Sergeant Strong, a tall Canadian in his forties who had a thick brown mustache and a runner's build and wore a dark beret with a gold crest. He had killed people in the Balkans, Afghanistan, and places he would not specify, and every time I pulled out my camera, he stepped out of view. He did not want me to use his real name. He was a patriot and a lifelong soldier, and recently he'd become a reporter for Canada's *Army News*. He roamed the ship with a pair of Nikons slung from his shoulders. We first met on the back deck, near the helicopter hangar, and he immediately asked who I thought owned the Northwest Passage. I said I wasn't sure. "It's ours," he told me. "It's fucking ours." Then he shared his solution for the territorial dispute over Hans Island. "We should just nuke Denmark," he said. He was kidding, of course. Canada has no nuclear weapons. His real solution was more typically Canadian, and it revealed him as a believer in the basic boots-on-the-ice premise of Operation Lancaster: If Canada backed up its Arctic claims with a physical presence, the world would recognize them. "Just put a trailer on the island," he said. "Two guys, two months at a time. Give them TVs and VCRs. And guess what: Problem solved."

The sergeant had a partner, Master Corporal Bradley, a giant videographer with whatever the opposite of a Napoleon complex is. Bradley's mustache was gray and waxed into dueling barbs, and he wore noise-canceling headphones even when he wasn't filming. He walked like a hunchback through the bowels of the *Montreal,* constantly hitting his head on doorways. The three of us, it turned out, would be part of the landing team forming Observation Post 1 on Devon Island. We would be joining eight Vandoos and four Canadian Rangers—Inuit reservists outfitted with red cotton hoodies—to go ashore at Dundas Harbor, a shallow fjord where the Royal Canadian Mounted Police had manned an outpost in the 1920s. Back then, two constables had been lost to self-inflicted gunshots to the head: the first, a suicide; the second, an apparent walrus-hunting mishap.

Two days before our "insertion," which is what everyone insisted on calling our mission to Devon, we were allowed to take a tour of the *Montreal*'s operations room—a cave of damp air lit only by radar and sonar screens and low red lights. Inside we met the ship's underwater-warfare officer. "Could you detect a passing submarine?" I asked. He could not. The ship couldn't drop sonar rays in the water without NATO permission. "They'd wonder why we were asking," he said. "And if we did detect something, we'd say, 'Hey, we found your sub,' and the Americans would say, 'No you didn't,' and we'd say, 'Yes we did.' It's a touchy subject." I asked about the relative size of the two navies. "The Americans, jeez, I can't count how many ships they have. They have sixty thousand people working in Norfolk alone. On one base. That's as many as we have in our entire armed forces. They have massive fleets. Massive. And we're obviously, you know, small." Our tour guide interjected, "But we can punch above our weight class." The officer agreed. "Yeah, we punch above our weight class."

One deck below the ops room was the lower-ranks mess, and I went there one afternoon to hear Commissioner Ann Meekitjuk Hanson, the formal head of Nunavut, address the troops. She told them about her childhood speaking only Inuktitut, her forced relocation to Toronto for schooling, and her Canadianized life in journalism and politics. "I have to disabuse southerners of their igloo notions," she said, "and explain that there's more to us than drumming and throat singing." A sailor named Roberts, one of perhaps five black people on the entire ship, asked how climate change was affecting the Inuit way of life. The commissioner said that autumn was getting noticeably later, and that they were having difficulties forecasting weather and ice conditions; now there were only six seasons rather than the traditional eight. She showed us slides of her homeland and put a cassette into a boom box to play some throat-singing music for us.

After the music stopped, I walked down the hall and found Sergeant Strong once again promoting his plan for the Hans Island dispute with Denmark. "It could be something as simple as putting a couple of guys up there with a trailer," he told a reporter from one of the in-flight magazines. "How much would that cost? The problem would just go away."

THAT OCTOBER, I traveled to Vancouver to meet the legal scholar Michael Byers, the former director of Duke University's Canadian Studies Program and a widely respected expert on Canadian security and sovereignty. Byers, who was a young-looking forty years old and wore the same two days' worth of beard he seemed to display every day, had recently returned home, surrendering his U.S. green card to a border guard in a burst of patriotism. He had taken a position at the University of British Columbia, and I was invited to sit in on his graduate seminar on climate change, a ten-person class held in a corner room with tall windows looking out on tall fir trees. When I walked in fifteen minutes late, a lanky student named Ryder McKeown was delivering a Power-Point presentation called "Climate Change and National Security." He wore jeans and glasses and Puma sneakers that happened to be red, white, and blue.

"Given the choice between starving and raiding," one of McKeown's slides read, "people raid." He wasn't talking about refugees from the tropics—at least not just them. The United States has a worsening water shortage, he said, and Canada has 20 percent of the world's freshwater. He described "fantastic schemes" to export it across the border in bulk, including NAWAPA, the North American Water and Power Alliance, a 1960s proposal by the Los Angeles engineering firm Parsons to divert Canadian rivers to run southward rather than northward. In another plan, fjords in British Columbia would be dammed at one end and filled with freshwater; tankers would arrive, top up, and chug south to California. "We have it," he said. "They want it." Byers jumped in. "We are talking about 300 million people"—ten times the population of Canada—"with the world's largest military and with a desperate need for water," he said. "To some degree the constraints of international law will fade into the background. But luckily, water conservation is much cheaper than enormous engineering projects. They'll find it hard to justify the expense."

The discussion turned to the Northwest Passage, where the United States has twice enraged Canadian nationalists by sending ships through without asking permission. The 1969 voyage of the SS *Manhattan,* an ice-strengthened supertanker that tested the frozen route's viability for transporting North Slope oil (the verdict: not yet), led to 1970 legislation in the Canadian Parliament that asserted Ottawa's right to control Arctic traffic, which in turn led to failed eleventh-hour maneuvering to forestall the new law by Henry Kissinger and the U.S. State Department, then to a retaliatory cut in U.S. imports of Canadian oil. The 1985 crossing of the U.S. Coast Guard icebreaker *Polar Sea* led to more uproar and the negotiation of an informal "don't ask, don't tell" policy: Before making any transits of the passage, the coast guard now notifies Canada (without exactly asking); Canada agrees never to tell its neighbor no. American submarines already use the passage to travel between the Atlantic and the Pacific, and I had heard unprovable tales of Inuit hunters mistaking those subs for whales and shooting at them, only to have their bullets bounce off.

"We are talking about moving from a country that, in practical terms, had two coastlines, to one that now has three coastlines," Byers said. "And we're being told that our new third coastline isn't subject to full

Canadian jurisdiction—that it's the Wild, Wild West." He said that drug smuggling, gun smuggling, illegal immigration, and environmental damage could go unchecked if Canada didn't take control. McKeown suggested there was a deeper threat as well. As divisive as the Northwest Passage may be, he said, Canada and the United States are drawn together in times of crisis—not pulled apart. He rattled off examples of cross-border cooperation: the Permanent Joint Board on Defense in 1940, NATO in 1949, NORAD in 1958, the Smart Border Declaration in 2001. In the mid-1950s, when it mattered that the shortest flight path from the Soviet Union to the United States was over the Arctic, the fifty-eight radar posts constituting the Distant Early Warning Line were built with mostly American money on mostly Canadian land. If climate change is truly as disruptive as both world wars, might Canada be drawn into an inescapable embrace with America?

McKeown was running out of time, so he raced through his last slides, laying out a climate-change scenario designed to "stretch our way of thinking": First, rising seas flood Bangladesh, Mumbai, and Shanghai. Refugee applications then flood Canada. A terrorist group based in Canada soon attacks America. The United States closes its borders. In retaliation, Canada ceases water exports. But then, as immigrants sneak in from the Arctic and Russian and Chinese subs cruise the Northwest Passage, Canada asks for America's help. It offers unfettered access to its resources in return for security. "Canada," McKeown concluded, "remains an independent country in name only."

Byers let that sink in. "If we're in a *Mad Max* world, when things are increasingly dangerous and it's survival of the fittest," he said, "it's not implausible to argue that our future is bound to the United States." He was playing devil's advocate. It worked. The class erupted. "Integration is a slippery slope," said McKeown. A student on the far side of the room agreed. "We could lose our central-banking independence, our monetary independence, our social democratic Canadianism," he said. "Our sovereignty is us, right? Without it we lose independent policy all over the board."

"Has anyone here been to Puerto Rico?" Byers asked. "Is it part of the United States?" The students answered that it was a commonwealth, a protectorate. "They're American citizens—sort of—but they can't vote," one said. "They don't have minimum-wage standards," said another. "There are a lot of people who support greater integration with the United States," Byers concluded, "and they're all under the assumption that we would become the next California—that we would become a state. But someone once told me that we Canadians need to pay more attention to Puerto Rico."

I was reminded of a Canadian radio contest some years earlier in which listeners were challenged to come up with a national slogan equivalent to "as American as apple pie." The winning entry: "As Canadian as possible under the circumstances." The conversation was far from the bravado of Operation Lancaster, but it was a flip side of the same coin. Canada was maneuvering to become one of the winners in a warming world, but a separate and equal goal, which I would soon see mirrored across the planet, was to avoid becoming one of the losers.

OUR INSERTION ONTO Devon Island began with a frenzy of packing and map reading and sorting through food rations in the helicopter hangar. A rope ladder was soon thrown over the side of the *Montreal*, and we put on black life jackets and climbed down to a Zodiac raft that was pitching on six-foot seas. The Vandoos' sergeant went first. The surprisingly graceful Bradley, all three hundred or so pounds of him, came last. We filled the front of the Zodiac with rucksacks and ration packs and weapons, and then we zipped across the ocean until the *Moncton*, a small warship just shorter than a hockey rink and supposedly better than the *Montreal* at landings, appeared out of the haze. We scaled its ladder and formed a bucket brigade to unload the gear. The *Moncton* was homey—its crew consisted of forty reservists—and so tight on space that the Vandoos had to set up cots in the hallways. Most of its sailors were as new to the Arctic as we were.

I'd known Devon Island only as the site of NASA's Mars on Earth project, in which investigators attempted to live on a rocky, frigid, arid analogue to the red planet, and its beauty surprised me when we approached the next day. It loomed large even from thirty miles out, its glaciers pouring down from desolate three-thousand-foot peaks. The fog was gone, the sun was high, and icebergs kept floating past. The water was milky, glacial. An Aurora surveillance plane appeared and made a triple pass above us, plumes of smoke trailing behind its four props.

We sailed in from the east, and as we turned the corner into the fjord, we were surprised by a ship sneaking up from the west: the Russian-flagged, Australian-chartered, sixty-six-hundred-ton *Akademik Ioffe*. It was a tour boat. I recognized it from the watchman's picture book, in which its photograph was sandwiched between images of Danish warships and surveillance aircraft. The *Ioffe*'s ice pilot radioed over. "Good afternoon, Warship 708, this is the *Akademik Ioffe*. We are a small passenger ship, an expedition ship. We have many Canadians—myself included—on board." His voice had a slight tremor to it. "It looks like you'll be into Dundas Harbor before us, so we'll be sure to stay out of your way." The officers on the *Moncton* snickered and rolled their eyes, pleased at the fear they generated. "You're damn right," one said. "I can't believe he called us before we called him," said another. It was reminiscent of the confrontation with the *Killer Bee,* only this time with a real, albeit Canadian, foe.

Our warship surged past the tour boat and arced a dramatic right turn into the fjord. We then slowed to a crawl. Our fifty-year-old charts couldn't tell us how deep the harbor was, and the captain was worried that we might run aground. We took depth soundings and peered into the silty water. The charts said it was thirty feet deep. Our sonar said it was more than two hundred. Best to stay put. We dropped anchor a mile offshore and began the slow process of readying the Zodiacs. The *Akademik Ioffe* steamed past us and anchored a half mile closer. As the Vandoos put on orange survival suits and the crew of the *Moncton* put on baby-blue helmets, the *Ioffe* put boats in the water. "They're beating us!" someone yelled. A hundred tourists made it to shore before our dozen soldiers were off the ship.

The Canadian Forces reached the narrow, rocky beach just as the *Ioffe*'s tourists were finishing their stroll. The tourists were white-haired and frail and dressed in matching blue-and-yellow Gore-Tex jackets. They had paid as much as $8,845 apiece to see the vast, empty Arctic. Cameras and binoculars hung from their necks. They seemed confused. The Vandoos shouldered their hundred-pound packs and struggled mightily up the beach, grunting, assault rifles in hand. They proceeded across the soggy tundra, their boots sinking into the mud with every step. One of the guides from the *Ioffe*, a man from Seattle with a bushy beard and a brown fedora, interrupted one march to remind us to "leave no trace"—Devon's environment was fragile. Sergeant Strong looked at the American. "A lot of us have spent a lot of time in the north," he said. "We're actually here to protect it."

DOWN IN AMERICA, it was the tail end of the second Bush administration, and Arctic policy and climate policy had both been adrift, the former because national-security priorities had shifted (no longer did we fear a Soviet attack from the north), the latter because economic priorities had not (to cut carbon was to cut fossil fuels was to cut the engine of a century of meteoric growth). But when I made visits to Washington in the winter and spring of 2007, I found the beginnings of another shift. Now the Arctic was becoming an issue again—an economic issue—and climate change was becoming a national-security issue.

The first report to explore the link between climate and security was commissioned by the Pentagon and written by the futurist Peter Schwartz, the former head of scenario planning for Royal Dutch Shell, where he foresaw such world-shaking events as the collapse of the Soviet Union. His twenty-two-page *An Abrupt Climate Change Scenario and Its Implications for United States National Security* was released publicly in 2004, and among its recommendations were that the nation "ensure reliable access to food supply and water," "prepare for inevitable climate driven events such as massive migration, disease, and epidemics," and research means to reengineer the global climate to our liking, known as geoengineering. In one two-month period in the spring of 2007, Schwartz would publish a second report (for an unnamed agency), the Senate would direct the sixteen main spy agencies to help the National Intelligence Council produce its own classified report, and eleven prominent retired generals and admirals with the Center for Naval Analyses would release yet another landmark study, *National Security and the Threat of Climate Change*. Soon, such studies would be as frequent and predictable in Washington as the May rains, a reliable source of employment for think tankers, and most effectively said the same thing: While climate change was not an existential threat for America, not in the way it was for Kiribati or Bangladesh, it portended endless mop-up. The Pentagon's fears, Schwartz told me, could be boiled down to a single word: Mogadishu. "Massive drought led to famine, which led to the collapse of Somalia, which led to the UN intervention, which led to the U.S. intervention, which led to a military disaster," he said. "They see a string of Mogadishus rolling off into the future."

Five miles from the Pentagon, in a nondescript office tower in downtown Arlington, was the U.S. Arctic Research Commission. When I visited, its seven commissioners and three staffers were the nexus of the country's Arctic policy, insofar as it existed. Its executive director proudly showed me a new pan-Arctic bathymetric chart the commission had helped produce: The North Pole was at the center, and the contours of the oil-rich seabed—still the least mapped part of the planet—were revealed as never before.

I asked first about the Northwest Passage, but in America, unlike Canada, the topic elicits little passion. The United States does not disagree that the Northwest Passage runs through Canadian waters. Its claim is that the passage is an international strait like Malacca, Gibraltar, Bab-el-Mandeb, the Dardanelles, and the Bosporus—a waterway that should be open to container ships and oil tankers from all nations. The European Union shares America's interpretation, and China—another country with much to gain from an open passage—had recently signaled its thoughts when its 550-foot icebreaker *Snow Dragon* appeared in the Arctic and the captain nonchalantly landed passengers at the Canadian settlement of Tuktoyaktuk.

"Why should we be the ones negotiating with Canada over it?" exclaimed George Newton, a mustachioed former nuclear-submarine captain and longtime commissioner. "Shouldn't Japan, which has a large fleet, take the lead? Or Maersk, the big shipping company in Denmark? Shouldn't Denmark be in on this?" He explained that the right of "innocent passage" through active straits was enshrined in the UN Convention on the Law of the Sea—the so-called constitution of the oceans, a treaty signed at the latest count by 164 countries. (The United States drafted much of the law but, owing to conservative wariness of UN pacts, is not yet party to it.) Newton granted that legal arguments over the Northwest Passage are complicated by the ice: It is normally frozen over, so it's difficult to call it an active strait. And it's equally difficult to say it will never become one. The treaty's language regarding ice-covered areas hurt the U.S. position as much as the language regarding international straits helped it. Still, everyone in Washington was confident that the rhetoric could be toned down, that we could all act like adults about this. The economics of the thing made an agreement inevitable.

Ratification of the Law of the Sea treaty was and is the Arctic Research Commission's greatest goal. This has hardly anything to do with the

Northwest Passage, though, and almost everything to do with oil and gas. While the treaty establishes navigational rules and each nation's rights to the fish and minerals within two hundred nautical miles of its coastline, it also allows nations to claim further territory based on how far their continental shelves extend under the surface. "This is part of our land," the basic argument goes. "It just happens to be underwater." This provision, Article 76 of the Law of the Sea, could turn the world into a different place. In terms of seabed it could someday own, the United States, with its maritime borders extended and each of its island holdings surrounded by an oversize doughnut of sovereign ocean, would grow by 4.1 million square miles. It would surpass China, Canada, and Russia, with their own expanded holdings, to become the world's largest country.

In the shallow Arctic, nearly every patch of seabed could be claimed by someone, and America had a foothold in the form of Alaska. Article 76, the commissioners hoped, was what would help the country secure its share of polar petroleum—some $650 billion worth, by one count. It was the rule under which the five nations with Arctic Ocean frontage—Canada, Denmark, Norway, Russia, and the United States—would carve up the north. It was the terms of engagement for the last great imperial partition.

"Our need for oil is not going to go away," Newton said. "We're going to need every bit we can get our hands on. Even if we don't use as much in cars and trucks, we're going to need plastics, and fertilizers, and all these other things essential to daily life. The more oil that's tied to us directly, the more that comes in a pipeline or in a short trip in U.S. waters, the better off we are." Canada's Arctic archipelago, he said, was the next "oil elephant," and that's not all: An estimated twenty-one billion tons of coal sit on Ellesmere Island, and methane—another potential energy source and a greenhouse gas at least twenty times more damaging than carbon dioxide—bubbles up everywhere through the Arctic's melting permafrost.

The commissioners had watched Russia—which had just announced plans for a special army to guard oil rigs and pipelines—get rich off its northern petroleum fields. "Look at how that country is digging itself out of a quagmire," Newton said. "They're getting up, flexing their muscles, feeling like big boys—people gonna pay respect. That's all thanks to the Arctic."

Our campsite on Devon Island was a flat patch of high ground at the base of a reddish hill of scree. In front of us was the Northwest Passage; to our side was the fjord that was Dundas Harbor; and below us, a few hundred yards away, were the weather-beaten wooden buildings of the abandoned Mountie post. There were a few bergy bits in the bay, a few patches of yellow grass around the cabins. The greenest thing in sight was the Vandoos' lineup of A-frame tents, which had been set up one next to another in a very military row. Our crew of Inuit Rangers, two men and two women who had arrived earlier by helicopter, were zipped inside a nearby dome tent, playing cards. They kept bursting into laughter because one of them kept farting. I was wedged in my own green tent along with Sergeant Strong and Master Corporal Bradley, who had been told, incorrectly, that they would be provided shelter.

There was little to do. The first evening, the Vandoos' sergeant spent hours trying to make contact with the *Moncton* and the two observation posts across the sound. He received mostly static, even after his men restrung the wire antenna a few times. "Eeny stay-shun, eeny stay-shun?" he called into the void. "Thees ees Thirtee-one-Bravo . . . Say ageen? Say ageen? Say ageen?" One of the Inuit, who, unlike the Quebecer, had learned proper Canadian English in school, took over, with scarcely better luck. Four of the Vandoos, apparently bored, ran off and climbed the hill behind camp, returning only after their superiors yelled at them. I counted icebergs. There were fifteen in sight, including two ship-size giants and two halves of another iceberg that the Rangers had watched break apart earlier that day. The ice was moving, but so slowly that you had to look away for a moment to detect any change. The sun tried but never really set. Night at seventy-four degrees north, if you could call it night, was a three-hour period of darker-than-normal gray.

We learned that the electric fence we'd been allotted to fend off polar bears didn't work. Our bear protection consisted of two .303 shotguns and four Inuit, so the Rangers' sergeant had the entire camp huddle around his laptop to watch a bear-safety DVD. Advice: Poking it in the eyes won't work. He said that the .303's first three rounds would be "bear bangers"—firecracker-like shots meant to spook polar bears—and the last was a slug of lead. Under Canadian law, only Inuit were allowed to kill polar bears, and there were no exceptions. But if it became a matter of life or death, we should all be ready to use the lead slug anyway. We should aim for the bear's neck or just below its shoulder, where we would have a chance at piercing its heart.

More tourists invaded on the second morning—ninety-two of them, all in matching yellow Gore-Tex jackets—and streamed toward our tents. "It looks like the march of the penguins, eh?" said Sergeant Strong. An old woman with a beret and a tiny backpack arrived and fixed her gaze on him.

"Who are you?" she asked.

"Who are you?" asked the sergeant.

"I'm from the boat," she said. She was an American from Chicago. Her friend, who was Portuguese but lived near Canada in upstate New York, joined us. The conversation turned to sovereignty. "The Americans are essentially greedy," the friend said. "If there's oil up here, they'll be here. That's what all these wars are about." She mentioned Hans Island, and Sergeant Strong lit up.

"I think the simple solution for Hans Island is to just put someone there," he said. "You've just got to keep people there year-round."

The woman made a face. "But then the Danes would just send some-one there, too," she replied.

The pair walked off, and we stood there in camp, our hands in our pockets. "That woman has an opinion about everything," the sergeant said. "She was way out of whack," Bradley said. "It's just silly to talk about a fight over oil. We're producing so much and sending so much south that you guys don't need to come up here and take it—we're sell-ing it to you freely." We watched the tourists return to their ship in a flotilla of Zodiacs. As soon as they were back on board, one of the Rang-ers grabbed his rifle and shot a three-foot-long Arctic hare—a bullet between the eyes from two hundred yards away. He'd spied it earlier but hadn't wanted to scare our visitors. He skinned it and chopped it into chunks and left it on top of a plastic bag in the middle of camp, where it began to dry in the sun.

WE DIDN'T KNOW IT THEN, but only a year would pass before the Northwest Passage was ice-free and open for the first time in recorded history. The polar ice cap would break all records as it shrank to the shape of a kidney, sickly and thin, just eight hundred miles across at the waist. And Canada's show of force in the north, its grit, would be over-shadowed by an international incident courtesy of Russia. In August 2007, an icebreaker carrying the polar explorer Artur Chilingarov, the deputy speaker of Russia's parliament and a prominent figure in Vladi-mir Putin's United Russia party, found an opening in the ice above the North Pole. In went two submersibles, and for three hours they de-scended in complete darkness to the seabed fourteen thousand feet below. Then Chilingarov's sub used its mechanic arm to plant a tita-nium tricolor flag—the white, blue, and red of Russia—in the flat clay at the true North Pole and floated back up to the more than forty journal-ists waiting at the surface. "The Arctic," he proclaimed at the press conference, "has always been Russian." A meme—the Battle for the Arctic—was born.

For a year or two, everyone would pretend that Russia's flag, like Canada's sovereignty operations, possessed some sort of coherent geopolitical logic, that it fundamentally mattered. But the truth was, under the Law of the Sea, the partition of the Arctic was already well under way. It was just that it's less dramatic: What matter are bathymetric charts, seismic surveys, and good lawyers. Scientists are mapping the formerly uncharted seabed and they and politicians and lawyers will argue about what is and isn't continental shelf, and whose is whose, and then the warming Arctic will be split five ways by five rich countries whose historic emissions helped make it such a worthy conquest in the first place—no flags or warships needed.

Far from flouting international law in its rush north, Russia had been the first country in the world to make an Article 76 submission, in 2001. Its opening bid, which laid claim to 45 percent of the Arctic Ocean, was rejected at the UN because its data was incomplete, and it is now gathering more. One forty-five-day, sixty-scientist cruise took place two months before Chilingarov's North Pole dive, producing reams of data but zero media fanfare. In a dingy office down a backstreet in St. Petersburg, the geologists who have led this and other follow-up surveys would proudly show me photographs of their approach to gathering seismic data: men pushing a golf-cart-size mesh sack of dynamite into an ice opening.

Canada signed the Law of the Sea in 2003, Denmark in 2004. Despite remaining tensions over Hans Island, the two countries have worked together to try to prove that the Lomonosov Ridge, the eleven-hundred-mile, trans-Arctic, undersea mountain range that may justify Russia's claim to the North Pole, is also connected to Canada's Ellesmere Island and Denmark's Greenland. America has sent scientists and State Department representatives north of Alaska to prepare claims to parts of the Chukchi and Beaufort seas; I once spent a drama-free month with them on an icebreaker. As for the fifth Arctic Ocean nation, Norway, it made its Article 76 submission in 2006. Wielding data collected by its Ministry of Petroleum and Energy, it claimed 96,000 square miles of ocean floor, reserving the right to claim more once it and Russia resolved their disputed border in the oil-rich Barents Sea—which they quietly did four years later.

Only Russia would say out loud what the other Arctic nations were beginning to recognize. "Global warming is not as catastrophic for us as it might be for some other countries," declared a spokesman for the Ministry of Natural Resources. "If anything, we'll be even better off. More of Russia's territory will be freed up for agriculture and industry." Putin once put it more casually: "We shall save on fur coats and other warm things." In Moscow, I would hear warm words from Chilingarov himself. "Of course I'm for international cooperation in the Arctic!" he told me in his office in Russia's parliament, the Duma. He supported the Law of the Sea, and his explanation of the titanium flag was straight-forward: When an explorer goes someplace—be it the moon, be it Mount Everest, be it the true North Pole—he plants his country's flag. He signed a photograph of the flag and robotic arm for me, then stabbed dramatically at it with his index finger, pointing at the seabed. "Look here, and here, and here, and here," he said. "There is plenty of room for other nations' flags." If each country got a big enough piece of the Arctic, I eventually realized, there would be little incentive to fight over the scraps.

In time, I would learn that this was also the considered wisdom of America's spy agencies. In an amusingly furtive meeting in a Washington, D.C., Starbucks, I discussed the National Intelligence Council's classified report on climate change with someone who asked to be identified as "a senior intelligence official familiar with the issues." His perspective was clear. Be worried about refugees from the desiccating south, sure. Be worried about having to intervene in Africa's resource wars. Be worried about sea-level rise damaging key infrastructure at

home. But in the Arctic, when your country is one of just five laying
claim to a quarter of the world's remaining petroleum in a region others
consider the global commons, don't worry too much about the other
four.

AT THE OBSERVATION POST later that second day, the *Montreal*
appeared in the sound, with the smaller *Moncton* trailing it like a baby
elephant. They floated past. We observed them. The Vandoos' sergeant
fiddled with the radio and sang ballads in French-accented English:
"Are you lonesome tonight? . . . Are you sorry we drifted apart?" The
Inuit stared at him. "Elvees," he said. "You dunno Elvees?" Later, he led
a handful of soldiers on an unsuccessful fishing expedition. When they
returned, they stripped and dove into the Arctic Ocean, staying in long
enough to wash their hair with a bottle of Pert Plus shampoo.

We combed through the rations—unwanted items got thrown in
together in a cardboard box—and I goaded Sergeant Strong about the
previous day's fedora-wearing American, an invader who dared tell
Canadians how to treat Canadian soil. The sergeant was too Canadian
to enjoy the irony. "That's okay," he said. "He was right. We do need to
be careful about the environment up here." His earnestness affected me.
Oil tankers were inevitably on their way, and there is no proven method
to clean up a spill in ice. Crude gets trapped under it, retention booms
are difficult to deploy, and chemical dispersants fail in cold temperatures; one of the most successful methods has been to simply light the
oil on fire and watch it burn away.

The radio started working, and from across the sound we heard
accounts of what had happened to the other observation posts. The men
and women of Observation Post 2 had landed in heavy seas—the navy
boatmen had ignored the Rangers' recommendation to use an easier
beach—and two Zodiacs were swamped by waves. The soldiers had to

use their helmets to bail out the boats. Some returned to the *Montreal* just to get warm, while the rest made a rough camp at the bottom of a steep slope. In the morning they learned that their goal, an abandoned scientific research station, was still six kilometers away. They had to be flown there in helicopters. They arrived to find one of the air force's Twin Otters, which was meant to be running support, stuck in the mud of an airstrip. Their satellite phone couldn't get a signal, and their radio barely worked. At Observation Post 3, troops were almost eaten by a polar bear. Waiting for a helicopter lift, they had taken down their bear fence and removed the ammunition from their shotguns, standard procedure before a flight. The bear was slinking up a ravine, and it was fifty yards away when the helicopter pilot spotted it. He had to dive-bomb it with the chopper to scare it off. Compared with the monotony of Devon Island, this all sounded rather appealing. But it was our job to observe, and that we did. It was our job to be a presence, and that we were.

A fog rolled in again, and the world became spectral and gray. When it had passed, Sergeant Strong and I explored the Mountie outpost together. The front door was coated with faded red paint. Inside we found a sewing machine, a rusted fuel drum, and a wooden table stacked with books: *Two Black Sheep, The Astounding Crime on Torrington Road, Buck Rogers in the 25th Century*. On the wall, someone had posted an inventory from the summer of 1945: two dog corrals, one flagpole, one fire shovel, one kitchen table, four kitchen chairs, one coal heater, one forty-five-gallon water barrel, two blubber tanks. The graves of the two Mounties who died here were just up the hill. "If I had a warm cabin to come back to," the sergeant said, "I could do it. I could do a winter here."

We had three days to go. We built a fire. We stayed up later and later. The time passed without tick marks. One night, I stood alone outside my tent, looking out at the sun that never set. The two youngest

Vandoos—a sixteen-year-old and a seventeen-year-old—had been given the first watch. I saw one take out his video camera and start walking around the tundra, filming very little at all. His partner sat facing the Northwest Passage, raising his rifle and pointing it into space, then lowering it, then raising it, then lowering it.

TWO

SHELL GAMES

WHEN AN OIL COMPANY BELIEVES
IN CLIMATE CHANGE

M ore than thirty years ago, before Royal Dutch Shell became the first of the oil super-majors to take advantage of the melting Arctic Ocean, the founding father of its celebrated team of futurists made a visit to Japan. Pierre Wack was a thin, incense-burning Frenchman and a follower of the bald Greco-Armenian guru Georges Gurdjieff, who taught his pupils how to transcend the hypnotic state of "waking sleep" that he believed most humans occupied for the entirety of their lives. Wack himself spoke mostly in parables, and each year he took a few weeks of vacation from Shell to meditate with a second guru in India. On this trip to Japan, in the science hub of Tsukuba, north of Tokyo, he saw an art exhibit that he considered the perfect metaphor for his work at the oil company. The exhibit's various video screens simulated how different animals saw the world. The bee saw hundreds of tiny images, the frog a two-dimensional reality with no depth of field. The important one was the horse. Because a horse's eyes are mounted on the two sides of its head, the video screen showed the opposite of normal human perception. "Humans see peripheral objects, at the corners of our eyes, as blurred and distorted," explains Wack's protégé Peter Schwartz in *The Art of the Long View,* his book on the strategic tool Shell calls scenario planning. "We see the center in sharp focus. Horses, at least according to this Japanese representation, see the peripheral as sharp." The goal of the scenario planner was to be like the horse—looking at reality but looking especially at its fringes, where the surprises emerge.

In its simplest description, a scenario—a tool now adopted by everyone from Disney to the National Intelligence Council that has guided most of Shell's major decisions for a generation—is a story. Each scenario is a plausible story about a plausible future, researched and told by a futurist like Wack. Because stories are how humans mentally and emotionally understand the world, the act of imagining a scenario is thought to force decision makers to prepare for it. A scenario is not a forecast: Forecasts tend to assume that the future will be a continuation of the present, said Wack, and they are useless just when they are most crucial—when companies need to anticipate "major shifts in the business environment that make whole strategies obsolete." Wack's goal was to develop multiple versions of the future—an improvement on the scenario-planning technique he first learned from Herman Kahn, the pear-shaped contemplator of nuclear apocalypse at the Rand Corporation and Hudson Institute and perhaps the first person to call himself a futurist. "Herman Kahn was trying to get the future right—trying to get the one that was close to reality," Schwartz told me. "Pierre was trying to influence the decisions we make today." Rather than bet everything on a certain outcome, Wack had his oil company preparing to thrive no matter which version of the future came to pass.

In buttoned-down Shell, a conglomerate of sensible Brits and sensible Dutch who would grow the company to be by some measures the biggest in the world—eighty-seven thousand people in more than seventy countries and territories—the scenario planners were eccentrics who had direct access to top executives. They came up with futures that were simply too heretical for the suits to come up with on their own. "I had the feeling of hunting in a pack of wolves," Wack told an interviewer before his death in 1997, "being the eyes of the pack, and sending signals back to the rest. Now if you see something serious, and the pack doesn't notice it, you'd better find out—are you in front?" In Wack's time, the scenario-planning team foresaw the twin Arab oil shocks of the 1970s— inconceivable to executives because oil prices had been so stable for so long—and Shell thrived when its competitors did not. Historically the least profitable of the Seven Sisters oil companies, it would after a decade become the most profitable, momentarily surpassing even Exxon.

During the later tenure of the bearded, electric Peter Schwartz, a trained aeronautical engineer who had studied Tibetan Buddhism, worked with the "transpersonal psychologist" Willis Harman, and befriended Peter Gabriel and the Grateful Dead, Shell bested all the supposed experts who said collapse of the Soviet Union was impossible.

Oil companies are future oriented by their very nature—it takes decades to conduct seismic surveys, secure leases, drill test wells, hit pay dirt, find partners, erect rigs, start production, and suck a reservoir dry—but at Shell futurism became part of corporate identity. Other scenarios contemplated the rise of Muslim extremism, the world's growing environmental awareness, and, before the (World Trade Organization) riots in Seattle, a backlash against globalization. The scenarios carried evocative names befitting a good story—the Rapids, Belle Époque, the Greening of Russia, Devolution, People Power, Business Class, Prism—and were characterized, above all, by their openness to ideas that made executives squirm. It is therefore not surprising that Shell faced up to what would seem the most squirm-inducing future of all for an oil company: Along with BP, it was first among the majors to publicly accept the science of climate change.

Schwartz had helped build a large-scale climate model in the late 1970s at his previous job at the Palo Alto think tank SRI International, which had invented not only the computer mouse but also VALS ("values and lifestyles"), a research methodology that advertisers used to target specific segments of the American public. By the time of his arrival at Shell in 1982, climate change and emissions were already part of the oil company's scenarios, and it seemed inevitable, he told me, "that we would decarbonize over time—for many reasons, climate among them." This was one reason Shell began moving aggressively into natural gas, which is less carbon intensive than oil.

In 1998, another onetime scenario planner, Jeroen van der Veer, who would soon become CEO, directed a formal, company-wide study of climate change's impacts on Shell's global business. The result was an in-house version of the Kyoto Protocol: a goal to reduce the company's own greenhouse-gas emissions by 10 percent by 2002, an internal cap-and-trade scheme, a shadow carbon price, and a commitment to evaluate projects on the basis of not only the profit they would make but the carbon they would emit. Under the cap-and-trade scheme, individual Shell units were given permits based on their past emissions, then encouraged to reduce those emissions and sell the permits to units that needed them more. The program soon failed because it was voluntary—only those units able to easily cut carbon signed up—and because the few that needed more permits simply went to Shell headquarters and asked for them and headquarters handed them out. But the commitment to calculating internal emissions remains, and the company easily achieved its 10 percent reduction by 2002, largely by stopping some of the methane flaring that has long lit up the skies above its refineries in Nigeria. If you count carbon the way Shell does, looking only at its own operations, it is now no more of an emitter than the Marshall or British Virgin Islands. If you include emissions from the products Shell extracts from the earth and sells to the world, on the other hand, it is more like Germany—responsible for at least 3 percent of humankind's annual greenhouse gases.

A decade after van der Veer's first climate study, Shell went a big step further. In 2008, it publicly released two scenarios describing the world up to 2050, Blueprints and Scramble, that explicitly warned of the dangers of climate change. They also foresaw a massive boom in global energy demand. For the first time, Shell declared that it had a preferred scenario: The greener, less emissions-intensive Blueprints offered the brighter future, for the company and the planet. Van der Veer gave interviews: It should be made expensive to emit carbon and other greenhouse gases, he declared. Global cap-and-trade agreements were urgently needed. Efficiency standards should be imposed. All this would require more government regulation. "People always think . . . the market will solve all of it," he said. "That of course is nonsense." But as time passed and governments continued to do little to regulate emissions, Shell, following Wack's prescription, prepared to thrive in whichever version of the future became reality. It would become a test case for how forward-looking businesses might choose to navigate climate change on their own.

SHELL FIRST ANNOUNCED Blueprints and Scramble the same week its chief Arctic strategist, Robert Jan Blaauw, was at a conference called Arctic Frontiers in Tromsø, Norway: six days of speeches, coastal cruises, banquets, dance shows, and concerts also attended by the supermajors ConocoPhillips and ExxonMobil and large national companies including Norway's Statoil, Russia's Gazprom and Lukoil, Italy's Eni, and France's Total. On the first night, the eve of northerly Tromsø's first sunlight of the year, I watched as hundreds of executives and dignitaries stood together and ate ceremonial *solbolle* pastries—sun balls—and were welcomed by a native Sami performer in a wool sweater. "Are you feeling volfy?" he asked the polyglot crowd, and then he cradled the microphone and began to howl.

The previous summer had shrunk Arctic sea ice by an extra 500,000 square miles, twice the area of Texas, a dramatic record. Carbon emissions, most of them from fossil fuels, were largely to blame. Yet inside the auditorium, the question was not if the Arctic should now be developed but how. Norway's petroleum and energy minister: "I think it's important to recognize that this [melt] is also an opportunity." A fellow at Alaska's Institute of the North: "Also in the Arctic are 25 percent of the world's known coal reserves." A Norwegian social scientist: "The Arctic Climate Impact Assessment looks at how climate change impacts the conditions for oil and gas production, not how oil and gas production impacts global climate change." A ConocoPhillips executive: "Listing the polar bear under the Endangered Species Act will do nothing to shorten ice retreat." An author of the new Arctic Council Oil and Gas Assessment: "In general, animals that have feathers and fur have sensitivities to oil spills." A representative of Sarah Palin's office whom President Obama would later choose to head the Alaska Gas Pipeline Project: "As a noted historian said, 'The wealth generated by Prudhoe Bay and the other fields on the North Slope since 1977 is worth more than all the fish ever caught, all the furs ever trapped, all the trees chopped down; throw in all the copper, whalebone, natural gas, tin, silver, platinum, and anything else ever extracted from Alaska, too. The balance sheet of Alaskan history is simple: One Prudhoe Bay is worth more in real dollars than everything that has been dug out, cut down, caught or killed in Alaska since the beginning of time.' Yes, oil is everything to us."

At the end of the second morning, during a session titled "Environmental Challenges—Risk Management and Technological Solutions," the head of the World Wildlife Fund's Arctic program took the stage. "Apologies to the translators," he said, "because I'm going to talk very fast. Okay. Would you buy a house on a floodplain below the one-in-fifty-year flood level? Would you, if you were a regulator, set the standard for nuclear power stations for one serious accident every one hundred years?" His voice was loud, his accent Australian, and his gray mustache flapped up and down as his head whipped back and forth.

"I think the answer for most people here would be no," he continued. "So why do we, the most inventive and intelligent species on the planet, continue to undertake activities which have far more risk than any of the things I've just outlined?" He began reciting facts: the pace of Arctic warming, the contraction of Greenland's ice sheet, and the magnified sea-level rise from Siberia's swollen rivers. Carbon in the atmosphere was increasing by 1.9 parts per million each year, up from 1.5 parts per million during the previous thirty years. Natural carbon sinks—oceans, plants—could now hold 10 percent less carbon than fifty years ago, their efficiency as buffers weakening. "Since 2000, the growth in carbon emissions from fossil fuels has tripled—tripled compared to the 1990s!" he said. "We're exceeding even the highest IPCC emission scenarios." He showed us a graph and how we were above the red line.

"Okay, let's summarize," he said. "We lost 22 percent of the area of sea ice in two years. And you think that's business as usual. We've lost 80 percent of the volume of Arctic sea ice in the last four years, and we pretend this is just part of what we do." He looked at the crowd, visibly seething. "So, what can we conclude?" he asked. "That the emperor has no clothes. The expansion of oil and gas activities in the Arctic will fuel further greenhouse-gas emissions, which will, in turn, cause further warming and systemic changes to the Earth's system, which, in turn, will cause massive impacts in the Arctic and globally, which will hurt you and me. Ladies and gentlemen, we are living inside the paradox." He then called for a moratorium on all offshore oil and gas development in the Arctic. For a moment, there was a stunned silence—his anger seemed out of place, rude and unhinged, an interruption to what had so far been a pretty nice conference—then the crowd clapped politely.

In time, Shell would have a practiced response to the apparent paradox, which was to say that it was no paradox at all. "There was a question about whether there was a paradox to produce oil and gas in the Arctic," Blaauw told another conference. "I don't think so. The story is pretty simple. Today, we share the planet with 6.9 billion people. By 2050, there will be 9 billion people. In order to meet rapidly growing demand, especially from China and India, diverse sources of energy need to be developed in parallel. Renewable energy, yes, in ever greater volumes. We need to reduce CO_2 emissions. But also fossil fuels and nuclear. We need them all. As conventional oil and gas sources run out, we need to look at unconventional resources and unconventional loca-tions. And that's exactly where the Arctic comes in."

Shell executives also studiously avoid any suggestion that the com-pany's Arctic ambitions are tied to melting sea ice. This is, in many ways, reasonable. As Blaauw has pointed out, the last time oil prices were this high, after the 1970s oil shocks, Shell began exploring the Arc-tic's Chukchi and Beaufort seas, west and north of Alaska, only to give up a dozen prospects after oil prices bottomed out again. Its main Arctic drill ship, the twenty-eight-thousand-ton, 270-foot-diameter *Kulluk*, which it purchased in 2005 and began refurbishing at an eventual cost of more than $300 million, is more than thirty years old, having been built during this same period of high prices. Now prices were back, global supplies were further depleted, and improved technology also made the Arctic more commercially viable: Thanks to a new explora-tion technology, directional drilling, no longer does every well require an expensive, environmentally disruptive platform of its own. Rather than dozens of pinpricks in the seabed, a single platform could drill a web of wells in every direction. There was another key change: In the 1980s and 1990s, the eight-hundred-mile Trans-Alaska Pipeline was full of Prudhoe Bay oil, too full to carry anything more. Now it was running so dry that Alaska officials were desperate to find new supplies, lest the depleted oil in the pipes get so cold that it began freezing in place.

But melting ice is not irrelevant. In public statements and private conversations, Shell officials acknowledge the following truths: Climate change is real. Climate change is melting the Arctic. Ice-free seas are easier for shipping. Ice-free seas are easier for spill cleanup. Ice-free seas are easier for seismic survey, which, as the company Web site explains, "enables explorers to see through solid matter in the same way an ultra-sound can see a baby inside its mother." And in places like Alaska, gov-ernments allow oil drilling only during the ice-free summer, a season that is lengthening year after year. As a Shell vice president once told a crowd at yet another conference, "I will be one of those persons most cheering for an endless summer in Alaska."

I caught up with Blaauw in the Arctic Frontiers coffee line one morning, and he voiced a similar thought. An eighteen-ship Shell armada had recently sailed to the Beaufort Sea, where the company had again acquired drilling leases in 2005, but it was blocked in court by a coalition of Native and environmental groups before summer exploration could begin. "It was such an abnormally low ice year," Blaauw told me, "so it's a pity we weren't allowed to go forward with the drilling." The Arctic wasn't like Saudi Arabia. "If you lose your chance to drill in the Middle East," he said, "you can go back in six weeks. But the Arctic is slow, very slow. You must wait an entire year until the ice is gone again." I asked if talk of the high north as the world's next oil elephant was overblown. It was not, he said. I should keep an eye on Alaska's upcoming Lease Sale 193, the first offshore auction in the Chukchi Sea in seventeen years. "There are enormous hopes for the Arctic," he said, "and I think you'll see that reflected in the prices at this Anchorage lease sale."

IN THE OPTIMISTIC WORLD of Blueprints, wrote Jeroen van der Veer in the booklet introducing Shell's two scenarios to 2050, "growing local actions begin to address the challenges of economic development, energy security and environmental pollution. A price is applied to a critical mass of emissions giving a huge stimulus to the development of clean energy technologies." There would be energy efficiency measures, electric cars, solar panels—"increasingly a world of electrons rather than molecules"—and, crucially, widespread adoption of carbon capture and storage, or CCS, the still embryonic process to catch carbon at power plants before it enters the atmosphere. CCS would keep greenhouse gases in the ground, and it would keep fossil fuel companies in business. Shell would prepare for either scenario, van der Veer wrote. "But in our view, the Blueprints' outcomes offer the best hope."

Blueprints, along with its counterpart, Scramble, was imagined as a plausible outcome of what Shell called the three hard truths: There would be a step change in global energy use. ("Developing nations, including population giants China and India, are entering their most energy-intensive phase of economic growth.") There would not be enough conventional energy to keep up. ("By 2015, growth in the production of easily accessible oil and gas will not match the projected rate of demand growth.") And climate change and other environmental stresses were real and getting worse. ("People are beginning to realise that energy use can both nourish and threaten what they value most—their health, their community and their environment, the future of their children, and the planet itself.")

Change comes from the bottom up in Blueprints as people's fears about their economy and quality of life lead to local action, which leads to regional, national, and eventually international action—"a critical mass of parallel responses to supply, demand, and climate stresses." Carbon trading accelerates, the story goes, "and CO_2 prices strengthen early. Perceptions begin to shift about the dilemma that continued economic growth contributes to climate change." Even in the developing world, "people make the connection between irregular local climate behaviour and the broader implications of climate change, including the threats to water supplies and coastal regions. After the Kyoto Protocol expires in 2012, a meaningful international carbon-trading framework with robust verification and accreditation emerges from the patchwork of regional and city-city schemes." By 2050, under Blueprints, the vast majority of coal- and gas-fired power plants in the world's richest countries would have CCS, reducing overall emissions by up to 20 percent.

Jeremy Bentham, the British theater enthusiast and former head of Shell Hydrogen who now sits in Wack's and Schwartz's seat leading the scenarios team, later explained to me why having a carbon price would be crucial to having CCS. "A rule of thumb is that a one-gigawatt coal-fired power plant costs $1 billion," he said, "and it's another billion to equip it with CCS. There's no return on that second billion unless you have carbon dioxide pricing." There was another rule of thumb to keep in mind. "Once something is technically and commercially proven," he said, "it will then grow at double-digit figures." But 25 percent growth per year, projected over thirty years, was still minuscule. "That's just 1 percent of the global energy system," he continued, "because the global energy system is so large." In other words, Blueprints, Shell's hopeful scenario, was only so hopeful. "The best climate outlook, pushed to extremes of plausibility, was Blueprints," he told me. "Blueprints was a 3.5-degree kind of outlook. I think we can be open about the fact that we hope for something more, but we have to think about what it's like to operate in a world that's on that trajectory. Ocean-level rises. Climatic turbulence, storms, and whatnot. I used to be a physicist. The more energy you're capturing in any fluid, the more turbulent would be the behavior."

The companion to Blueprints, Scramble, described an even scarier future—and one that has seemed closer to reality in the half decade since. Its world's key feature is that it is reactive: "Events outpace actions." Countries keep burning coal and oil deposits, racing one another for them, emitting more and more carbon, and changing course only when nature forces them to. "Policymakers pay little attention to more efficient energy use until supplies are tight," wrote van der Veer. "Likewise, greenhouse gas emissions are not seriously addressed until there are major climate shocks."

In the early years of Scramble, despite some "turbulence," the global economy continues to grow. "National governments, the principal actors in Scramble," Bentham's team explained, "focus their energy policies on supply levers because curbing the growth of energy demand—and hence economic growth—is simply too unpopular for politicians to undertake." Much of the energy powering these unfettered times comes from coal, the dirtiest fossil fuel, which emits twice as much carbon as does gas and nearly a third more than does oil: "Partly in response to public pressures for 'energy independence,' and partly because coal provides a local source of employment, government policies in several of the largest economies encourage this indigenous resource. Between 2000 and 2025, the global coal industry doubles in size, and by 2050 it is two and a half times as large."

In their hunger for energy, the nations of Scramble also turn to biofuels. These compete with agricultural production, especially in the corn-growing regions of the world, and drive up global food prices. Biofuels importers inadvertently encourage poorer nations to destroy rain forests in order to grow palm oil or sugarcane, resulting in major emissions of the CO_2 stored in the soils of the former forests. Investors also pour "more and more capital into unconventional oil projects"—such as Canada's tar sands—that are opposed by environmental groups for their high emissions and water use.

In Scramble, climate campaigners get louder, but "alarm fatigue afflicts the general public. International discussion on climate change becomes bogged down in an ideological 'dialogue of the deaf' between the conflicting positions of rich, industrialised countries versus poorer, developing nations—a paralysis that allows emissions of atmospheric CO_2 to grow relentlessly." Toward the end of the scenario, when the supply crunch and climate change are impossible to ignore, emissions begin to level off. But CO_2 concentration is heading above 550 parts per million—200 more than the red line of 350 parts per million identified by campaigners and many diplomats and scientists. "An increasing fraction of economic activity and innovation," the scenario planners write, "is ultimately directed towards preparing for the impact of climate change." That is, the world must adapt to what it has become.

When I asked Bentham in 2012 if the future was looking more like Scramble than Blueprints, he was uncharacteristically concise. "Yeah," he said. "That's the view."

BEFORE I FOLLOWED Shell to Alaska for the lease sale, I traveled up the Norwegian coast from the conference in Tromsø for a glimpse of the future Arctic. The formerly dingy fishing town of Hammerfest was home to Snøhvit, or Snow White, the world's northernmost liquid natural gas operation, watched closely by Shell and its rivals. It was the day before the planned start of production when I arrived, and the $10 billion installation had long ago taken over a once grassy island abutting town. Viewed from Hammerfest's newly glitzy shopping mall, it was a tangle of smokestacks, lights, and tubes, backed by a fjord and a row of snowy peaks. The gas field was farther offshore, in the Barents Sea, eight hundred feet underwater and connected to the island by eighty-nine miles of pipes. Production was behind schedule. A few months earlier, the winds had shifted as engineers were putting the plant through the paces, and its flares—chimneys burning off excess gas—coated cars and homes in a layer of black soot. The plant operator, Statoil, Norway's national petroleum company and Shell's soon-to-be rival in the Alaska lease sale, brought in doctors to test for carcinogens and community liaisons to hand out reparations checks.

Here at the top of Scandinavia, where the North Atlantic Current left the coastline mostly ice-free, the Norwegian national schizophrenia was amplified. The second-richest country in the world with the second-largest sovereign wealth fund in the world, a $500 billion reserve known colloquially as the Oljefondet, or Oil Fund, Norway was flush enough from offshore petroleum that it could afford to be concerned about the environment: In 2000, it became the world's first and so far only country to sack its government over lack of progress on carbon emissions. It was serious about the Kyoto Protocol, so much so that Snøhvit would eventually become a CCS test facility, thus a test of whether a scenario like Blueprints could ever come to pass. It would reinject CO_2 into the seabed after sucking out all the natural gas. In the meantime, Snøhvit's production problems might single-handedly cause Norway to miss its Kyoto targets. And the country's sovereign wealth fund, which on ethical grounds excluded investments in tobacco companies and arms dealers, counted Shell—perhaps Norway's equal in schizophrenia—as its single biggest stock holding.

As Hammerfest waited for the plant to fire up again, I had a tour of the island with a Statoil spokesman, clearing security, driving through a tunnel beneath the fjord, and passing barracks of imported workers: Turks, Greeks, Slovenians, Poles, Finns, and Russians. The wind was blowing again, and the *Arctic Princess,* one of the world's largest natural gas tankers, was anchored in the bay. But what interested me most was the Faustian bargain back in town. In a pizza restaurant in the center, I met the only local politician opposed to the plant: a nineteen-year-old from the revolutionary-socialist Red Party. We mostly talked about shopping. "I love eBay!" she said. She told me she used it to order American clothes from four thousand miles away. The gas from Snøhvit would go to Bilbao, Spain, and, eventually, to Japan and China via the Northeast Passage, the newly passable shipping lane above Russia that is also known as the Northern Sea Route. Much of the money would stay here. Statoil paid ninety-four-hundred-person Hammerfest $22 million a year in property taxes, and that, the socialist admitted, bought loyalty. Even her mom was in favor of Snøhvit.

In his bay-front office, Hammerfest's deputy mayor touted his town's new projects: renovated primary schools, a bigger airport, a flashy sports arena, a "full-digital," glass-walled cultural center. Home prices had doubled in five years; strollers were everywhere in the snow-covered streets. It was easy to forget that until recently Hammerfest was a dying town, shrinking in population, the most violent place in Norway. "It was clean fighting, not so much with knives and such," he assured me. I asked about the soot from the flares. "People didn't like it," he said, "but they accepted it."

It was 2:00 p.m., the high north in winter, and it was becoming dark. I stepped out just in time to see Snøhvit come to life—the Arctic on fire. A flame spouted four hundred feet, five hundred feet from the tallest chimney, dwarfing the mountains, hanging high over the town, bathing it in orange light. From two miles away, I could hear it burn, and I could feel its heat on my face.

"I couldn't start without saying thank you," Randall Luthi told the crowd at Chukchi Lease Sale 193, and a sea of oil traders and lobbyists stared silently back at him, or perhaps past him, at the map of petroleum blocks projected on a floor-to-ceiling screen. "The thank-you goes to industry for making their interests known," he said. "But thank you also to those who have voiced concerns—because this is a time that is very indicative of the way the world is today, of the way our economy is today, of the way our energy future is today. These are tough times with tough decisions and tough questions. One question I've been asked: Why have this sale?"

Outside the hall at Anchorage's main public library, a group of activists, two Inupiat Eskimo men and three white women—one wearing a polar bear costume and a pair of Sorels—waved handwritten signs: "Oil and Polar Bears Don't Mix!" "Keep Big Oil OUT of Our Garden!" "Chill the Drills!" "Don't SpOIL My Dinner!" Their breath condensed in the frigid air. Inside, in front of the screen, three schoolmarmish staffers armed with tape, boxes of paper clips, and bottled water guarded a table covered with blue file folders: the bids. Luthi, a rancher from Freedom, Wyoming, whom George W. Bush had appointed director of the Minerals Management Service (MMS), wore an ill-fitting gray suit and stood at a podium emblazoned with the MMS seal. "Mineral Revenues–Offshore Minerals–Stewardship," it read, the words encircling a golden eagle. The MMS had yet to be rocked by its "oil for sex" scandal, yet to be blamed for lax oversight in the Deepwater Horizon catastrophe in the Gulf of Mexico, yet to be re-formed, or at least renamed, as BOEMRE, the Bureau of Ocean Energy Management, Regulation, and Enforcement, then as plain BOEM. Lease Sale 193, under which 45,900 square miles of Arctic seabed would be offered up in 9-square-mile chunks, was going forward despite serial delays by the MMS's parent agency, the Department of the Interior, on a closely watched decision: whether the polar bear, a resident of the Chukchi's retreating sea ice, should be listed as global warming's first official threatened species. It would be the most lucrative lease sale in the history of the Arctic Ocean, and Shell would scramble ahead of its rivals with high bids totaling $2.1 billion.

Before the floor was opened, Luthi tried to answer his own question. "Why? Our demand for energy is going to increase by approximately 1.1 percent a year over the next generation," he said. "U.S. production is not expected to keep pace. Now, it doesn't take too much to realize that when you're demanding more than you're producing, there's a shortfall. The Chukchi Sea is widely considered one of the last energy frontiers in America. Now, I don't believe we should look at it as the last frontier, but rather as a frontier of unlimited opportunity.

"We understand the importance of the Chukchi Sea to the people who live along it," he continued. "We consulted with the communities, including the Native villages of Point Hope, Point Lay, Wainwright, Barrow, and the Inyoopit . . . the Inupit . . . I'm sorry, Inupiat Community of the Arctic Slope."

Someone in the crowd laughed.

"Inupiat," Luthi said again. "I always get that wrong. I sat back there and practiced . . ."

Another man stood to read off the bids—667 of them, another record for the Arctic. "We've estimated that this might take four hours to go through," he said, and he reminded bidders that electronic funds needed to be in a U.S. Treasury account no later than 2:00 p.m. the next day.

The first winning conglomerate was Spain's Repsol, unopposed for $75,050. There was no shouting, no excitement. It was a silent auction, with all the bids made in advance. The man opened them; we just sat there. He read in a near monotone—"Block 7011. One bid. Repsol, $75,050. Block 7019. One bid. Repsol, $75,050. Block 6868. One bid. Shell Gulf of Mexico, $303,394"—and the schoolmarms passed the blue folders down the table, left to right. The hall was quiet but for coughing.

"Block 6154. One bid. ConocoPhillips, $125,110," the man announced. "Block 6155. Two bids. First bid: Shell Gulf of Mexico, $4,106,999. Second bid: ConocoPhillips, $251,625. Block 6515. One bid. Shell, $508,900." And so it went. Shell kept winning: One block for $4,105,958. Another for $14,300,435. Another for $31,005,358. Another—and this one got a murmur from the crowd—for $87,307,895. Then another for $105,304,581.

Two hours passed. We had a halfway break in the lobby, where the activists' polar bear costume now sat crumpled on a stone bench near the window, next to a trader chatting on her cell phone. The crowd funneled back inside, and the presenter droned on. He read out the numbers in full: "one hundred and five million, three hundred and four thousand, five hundred and eighty one." As the total entered the billions, I lost all sense of scale. Shell. Ten million, one hundred and one thousand, five hundred and fifty. Statoil. Two million, seven hundred and sixty-two thousand, six hundred and twenty-two. Shell. Ninety-six thousand, six hundred and three. Shell. Fifty-four million, one hundred and four thousand, eight hundred and fourteen. Shell. Six million, fifty-seven thousand, six hundred and seventy-nine. Shell. Three hundred and seven thousand, seven hundred and fifty. Shell. Shell. Shell. One hundred and one thousand, three hundred and thirty. Eighty-two thousand and eighty-eight. Twenty-four million, three hundred and seven thousand, six hundred and one.

ONE COULD ALREADY SEE the writing on the wall: Shell's shift from the greenest oil company to the oil company targeted by Greenpeace for its

Arctic dreams. Its shift from actively pushing for a climate-change bill in the U.S. Congress to quietly recognizing that the government would accomplish very little at all. Its acceptance that the future was starting to look like Scramble. A few months after Lease Sale 193, Shell gave up its 33 percent stake in the world's largest wind farm, the thousand-megawatt London Array. Within a year, it had dropped all new funding for wind, solar, and hydrogen energy. It began securing government funding in Stephen Harper's Canada to build Quest, a first-of-its-kind, $1.35 billion CCS facility at the Athabasca tar sands that would inject captured carbon into porous rock more than a mile underground. But also, controversially, Shell invested heavily in the carbon-spewing tar sands themselves. If it extracted them without functional CCS, activist groups alleged in a 2009 report, it would become the most carbon-intensive oil company in the world.

Jeremy Bentham's team of futurists moved on to their next set of scenarios, which explored the "stress nexus" between water, energy, and food—a vital topic in a world adapting to climate change. "Water is needed for almost all forms of energy production," Bentham wrote. "Energy is needed to treat and transport water; and both water and energy are needed to grow food. These are just a few of the linkages." Climate change related to all three, and all three—whether in the form of deforestation for food production or carbon-intensive desalination for drinking water production—related to climate change. "I'm an ex–refining man," Bentham told me. "Water, always. Heating water, runoff water—water has always been an important operational issue. Now it is very much a central strategic issue." Bentham's deputy, a former BBC journalist, added that the local nature of water stress made it more politically explosive than carbon emissions. "Wars have been fought over water for years," he said, "whereas it's hard to imagine a war from CO_2."

In 2012, I asked Bentham about Shell's flight from renewables, and he assured me that its pullback from the London Array and other renewable projects looked different from the inside. "As we focus our attention on sweet spots, what are they? There's the recognition that there are some things that you can do well," he said, "and some things you find that you can't add value to." Wind was about turbines and other infrastructure that Shell didn't itself build. Solar was about silicon, also not an area of expertise. "Shell didn't have much to add," he said. But the company was moving forward with its Brazilian biofuels—second-generation crops that did not compete with food—and with its Canadian CCS. "And Shell has stepped over the barrier from more than 50 percent oil to more than 50 percent gas—we're now a gas company," Bentham added. "Gas is a very Blueprints kind of fuel."

Peter Schwartz, who long ago retired from Shell to spread the gospel of scenario planning as a kind of business consultant, was blunter. How, I asked, did it all make sense with Shell's stated preference for Blueprints? "It doesn't," he said. Then he caught himself. "Actually, in some ways it does make sense," he said, "because renewable energy has been more like Scramble than Blueprints. I mean, look at the United States. Are we gonna continue the tax credits or are we not? Right now the wind credits are all up for grabs again. And you've got a cap and trade system kicking in in California but nowhere else. Huh! How do you do that? That's not a Blueprints kind of world. In the Arctic, we're definitely scrambling. We have no blueprint."

THREE

GREENLAND RISING

AN INDEPENDENCE MOVEMENT HEATS UP

When I arrived in Greenland, the secessionists had gathered in a community sports hall halfway up the island's west coast. Upernavik was a town of eleven hundred people at a latitude of seventy-three degrees on a treeless tundra six hundred miles north of the capital, Nuuk. From Devon Island, where I'd sat at the observation post with the Vandoos, it was about five hundred miles due east, across Baffin Bay. But Danish-developed Upernavik stood in great contrast with the emptiness of Devon. The town had a fish plant, a hillside cemetery with concrete graves that were covered with plastic flowers, a single paved street, and an unmarked liquor store in a converted shipping container. Its wooden houses were painted in beautiful primary colors. Its teenagers hung out in the streets, blasting hip-hop from their cell phones, and in the mornings those streets were lined with yellow bags of excrement waiting to be picked up by sanitation teams. Upernavik was, like the rest of Greenland, oddly, lopsidedly modern—Scandinavian by design but not always by disposition.

Greenland had been a colony of Denmark for three centuries, and now it was on the verge of an oil and mineral boom that could help it become something else: the first country in the world created by global warming. I'd come here to join the secessionists' road show—and to witness the moment that some of the supposed victims of climate change began cashing in on it. Greenland's was an extreme case of the dilemma facing many citizens of the developed world, many northerners: If climate change wouldn't much hurt them personally—if it might even help—why not embrace it?

The road show was led by the Office of Self-Governance, and it consisted of half a dozen Greenlandic politicians—men and women wearing jeans, fleece, and tennis shoes—and dozens of town-hall meetings. In the run-up to a referendum in November 2008, they were trying to reach nearly all of Greenland: fifty-seven thousand people spread out in fifty-seven villages and eighteen towns across an area of 836,000 square miles, three times the size of Texas and fifty times the size of mainland Denmark. There are almost no roads connecting the island's settlements; it has two stoplights, both in fifteen-thousand-person Nuuk. We traveled by prop plane, helicopter, motorboat, and foot.

In the sports hall, one of the politicians warmed up the crowd with a funny story about a whale. He'd been a policeman here in the 1990s. The story went something like this: The police chief gets a call from a citizen. The citizen is a fisherman. He's caught a whale. He doesn't know what he should do with this whale. The chief says to the citizen, "Put it in the boat. We'll take care of it tomorrow."

Put it in the boat! Take care of it tomorrow! The crowd, roughly sixty people, roared with laughter. Mininnguaq Kleist, the thirty-five-year-old head of the Office of Self-Governance and its principal PowerPoint presenter, doubled over. I pretended to laugh, too, but I had no idea what it meant.

When it was Minik's turn to speak, he paced in front of his slide show wearing a headset, looking more like a televangelist than a revolutionary. Townspeople were sitting in red chairs below the basketball hoop, and out the window, in the bay near the fish plant, was an iceberg. Up for vote in a few months, Minik explained, was not full independence but a sort of half step, which was being called "self-governance"—"Namminersorneq" in Greenlandic, "Selvstyre" in Danish. Denmark, which gives Greenland nearly $650 million a year in subsidies—more than $10,000 a person—had signed off on it.

Greenlandic would become the official language if the referendum passed; Greenlanders would be recognized as a distinct people under international law. "We will take over policing," Minik said, "and immigration, education, and the courts." In the waters off Greenland's northern coast, the U.S. Geological Survey had just identified the nineteenth richest of the world's five hundred known petroleum provinces: an untapped Gulf of Mexico in the North Atlantic. To our south, near Disko Bay, Greenland's first oil leases had just been sold to the likes of Chevron and ExxonMobil. Shell and partners would soon claim a lease a hundred miles from Upernavik, in Baffin Bay. Onshore, glaciers were pulling back to reveal massive deposits of zinc, gold, diamonds, and uranium. "Control of mineral and oil resources will also be taken over by Greenland," Minik said.

They planned to drill themselves free. Under the agreement, the island would split mineral revenues with Denmark after keeping the first $15 million. As revenues went up, the $650 million annual grant from Denmark would go down. Eventually, over five years or ten, over fifteen years or twenty, if it warmed enough, if they drilled enough, the grant would hit zero, and Greenland would declare independence. In chemistry, there is activation energy: Add heat, get a reaction. In Greenland, there was global warming: Add heat, get a revolution. But this was secession at the speed of climate change, a slow burn.

When Minik was done, the mayor of Upernavik, a thin man with a few missing teeth, stood to pose a question. "Some of the money goes back to Denmark?" he asked, surprised. An old woman in a black sweater was next: "If we earn all that money, how much of it will stay in Greenland?"

But the politicians were banking on more than mining. Valuable fish stocks—cod, herring, halibut, and haddock—were migrating into

Greenlandic waters, moving north as the oceans warmed. There was a
rush of disaster tourists, people coming to see the glaciers slide into the
sea: Cruise ship arrivals had jumped 250 percent in four years, and
shops sold postcards showing melting ice with the label "Climate
Change and Global Warming." The expansion of the South Greenland
agricultural season—already three weeks longer than it was in the early
1990s—meant potato farms and carrot gardens and more grass for more
sheep. There were plans for a new aluminum smelter—360,000 tons a
year, the biggest in the world—to be built by Alcoa and run on hydropower from the island's gushing rivers. A pair of ships had just finished
laying high-speed Internet cable across the Denmark Strait, connecting
Greenland to Iceland and onward to North America, and there were
plans for fields of server farms—warehouses of computer processors
working for Google or Cisco or Amazon—to take advantage of the
cheap electricity and high latitude. "They normally need a lot of air-
conditioning," Minik explained. There were even plans for the melting
ice itself: water exports. "The Greenland ice cap has an estimated volume of 1.7 million km³, the world's biggest water reservoir," boasted a
Web site set up by the Secretariat of Ice and Water. Investors could sell
"two million years of history in a bottle!"

IT ISN'T ANYONE in Greenland's fault, but the Maldives are probably
doomed. Tuvalu is probably doomed. Kiribati is probably doomed. The
Marshall Islands are probably doomed. The Seychelles are probably
doomed. The Bahamas are probably doomed. The Carterets are probably doomed. Bangladesh, at least a fifth of it, is probably doomed. Large
portions of Manila, Alexandria, Lagos, Karachi, Kolkata, Jakarta,
Dakar, Rio, Miami, and Ho Chi Minh City are probably doomed. Water
enough to flood them all is stored in the world's biggest reservoir, the
Greenland ice cap, the frozen inland mass that covers 81 percent of the
island. The rate of the ice cap's melt had been increasing by 7 percent a

year since 1996. If someday it melts entirely, global sea levels will jump more than twenty feet.

In Alaska, villages such as Newtok, Shishmaref, and Kivalina were also endangered, made increasingly unlivable by coastal erosion, melting permafrost, and creeping salinity. Newtok's elders had acquired a new town site on a new island to their south—one with a hill—and were lobbying the state and federal government for the $130 million to move their 315 residents to it. The Shishmaref Erosion and Relocation Coalition had chosen a new site on the mainland, a few miles away, and was also awaiting funds. The residents of Kivalina were the plaintiffs in a conspiracy case against eight energy companies, including ConocoPhillips, ExxonMobil, and Chevron, who were accused of stirring up climate-change skepticism so they could produce more oil. The case—which would eventually be dismissed by a California judge—was being watched closely by American trial lawyers. As I heard one say excitedly at a conference, "This could really open up the floodgates . . . oh, probably not the right phrase."

Small island nations also considered climate lawsuits—in 2002, Tuvalu had threatened to file one against Australia and the United States—but mostly they, too, were looking for someplace new: Seventy-five Tuvaluans and seventy-five Kiribatians a year go to New Zealand under immigration quotas; the first five of seventeen hundred Carteret Islanders moved to newly purchased land in Bougainville, also in Papua New Guinea, in 2009. In Australia, a Tuvaluan-born scientist named Don Kennedy was drumming up support to buy an island from Fiji for his people. In the Maldives, the charismatic president, Mohamed Nasheed, the "Obama of the Maldives" who became the face of climate change before he was forced out in a coup, declared that he was looking to buy land in Sri Lanka or Australia, just in case. This prompted an Indonesian minister to announce that his country had some spare islands to sell.

In the Alps, melting glaciers around the Matterhorn had shifted a

border in place since 1861—it followed a ridge of snow that was no longer there—causing Italy and Switzerland to sit down and start negotiating a new one. In Kashmir, experts worried, the accelerating melt of the
Siachen Glacier would further provoke the India-Pakistan fight. The
world's shifting political map scared just about everyone. The NGO
Christian Aid estimated that by 2050, a billion people would be pushed
from their homes by global warming. Friends of the Earth said there
would be as many as 100 million such climate refugees. The IPCC said
150 million. *The Stern Review,* 200 million. The International Committee of the Red Cross said there were 25 to 50 million already.

The implications of the melt were so bad for so much of the world that
it seemed almost rude, even in Upernavik, to consider how good they
were for Greenland. Since 2003, the ice cap had shrunk by more than a
million tons, so much that the underlying bedrock rose four centimeters
each year, like a ship slowly unweighted of its cargo. In Greenland, the
land was rising faster than the sea.

MINIK LIKED TO GRAPPLE honestly with consequences. On the way
to Upernavik, in the Kangerlussuaq Airport, a building on the tundra of
western Greenland that felt like a ski lodge in the Alps—lounge chairs,
huge windows, a cafeteria with trays, rich tourists in Gore-Tex—I heard
his life story. He told me that after becoming Greenland's national bad-
minton champion, he'd gotten his master's in ethical philosophy at
Denmark's Aarhus University. His thesis, "Greenlandic Autonomy or
Secession: Philosophical Considerations," was a work of secession theory, the study of whether one country has, or doesn't have, the moral
right to break free from another. One early revelation, he said, had come
from what he called "my first philosophical crisis," after he'd tried to
apply the Aristotelian ideal of the good life to every little thing in his
real life: not every action can be moral.

He was straightforward about the life being wiped out by climate change. "It's a problem for hunters," he said. "Dog sledges fall through the ice. Or there is no ice." It was harder to get seals. It was harder to go ice fishing. In the north, it was hard to do anything but move to the larger towns.

Minik was equally straightforward about the Danes: They'd been mostly benevolent colonizers. In his thesis, and later in the self-governance talks, he'd used their own moral arguments against them. In only one place did his philosophy significantly break with that of the father of modern secession theory, Duke University's Allen Buchanan. "According to him, you have to be wronged to justify it," Minik told me. "Denmark has to wrong Greenland in a really bad way before we break away. I don't agree with that part. Sometimes you have to view this as a marriage: adults, consenting people, divorcing of their own free will."

Greenland had become part of Denmark in 1721, when the Lutheran missionary Hans Egede showed up and started saving souls. The first Danes taught the Inuit that hell was very hot rather than very cold, as they had previously believed. The Danes taught that communal living—shared food, shared hunting trips, shared wives—was sinful. They taught that rocks and birds were not endowed with spirits. Greenlanders had no bread or concept of bread, so Egede had translated another pillar of Western belief—the Lord's Prayer—to fit Greenlandic reality. "Give us this day our daily harbor seal," the Inuit prayed.

In the Danish colony, the crown had declared as early as 1782 that the Greenlanders' welfare should "receive the highest possible consideration, [overriding] when necessary the interests of trade itself." The Danes had harvested whales and fish and some coal, but they gave back homes and schools and hospitals. In 1953, they gave full Danish citizenship to every Greenlander. They gave students like Minik a free education at the university of their choice in Europe or North America. In Canada in the 1940s, meanwhile, Inuit were given numbered dog-tag-like IDs because they had no surnames, and they were moved to barren islands to reinforce sovereignty claims. In the United States, Inuit, including a boy named Minik, were put on display in the American Museum of Natural History by Admiral Robert Peary, the explorer who later claimed, probably falsely, to be the first to reach the North Pole.

Upernavik was seasonally frozen out of all ship traffic, possessed of no trees to cut or road links to deliver supplies, and two thousand miles away from Denmark. Yet it had this: a beautiful gym with a digital scoreboard, a hundred-foot-high ceiling, and long wooden beams five feet thick. The local hospital was staffed by Danes and Swedes, the Pisiffik supermarket price-subsidized, the cell phone signal strong, the street paved—not the mud tracks I'd seen in Inuit towns in Canada and Alaska. Nearby, a mountaintop had been lopped off, turned into a mesa: Upernavik's airport, its link to the world. The airport had a handicapped-accessible toilet.

From up here, the Danes, who got a fifth of their electricity from wind power, who just agreed to give up 98 percent of their territory, who would soon host the Copenhagen climate conference—dubbed "Hopenhagen" until the talks let everyone down—seemed oddly idealistic. Easy marks. I had to wonder at their motives. Did they plan to keep Hans Island? Did they plan to keep the Arctic Ocean seabed claimed under the Law of the Sea? But Minik didn't wonder, and he didn't much wonder if he was scorning the wrong people.

The day after the sports hall meeting, the politicians set out for the tiny whaling village of Kangersuatsiaq, where they would sit in a red community center and discuss the referendum with another audience. Minik and I followed in a twenty-two-foot fishing boat piloted by Upernavik's gap-toothed mayor, who pointed out a halibut trawler and told us that in the 1970s waters here were ice covered from late December until May; now they were open year-round. Shrimp trawlers had started coming to Upernavik, following their prey north, and so had herring—a fish that had freaked locals out when they first caught them. He told me about a new contest run by the Greenland Bureau of Mines and Petroleum: Send in rocks from your community. "If your sample is the best," he said, "you win 125,000 Danish kroner!"—more than $20,000.

We motored inland, cutting through waves at twenty-five knots, moving closer to the ice cap, and the temperature dropped ten degrees. We skirted a sheer, three-thousand-foot-high cliff of dark basalt that dropped straight into the fjord, staying away from its base to avoid rockfall.

Minik, bundled in a black Arc'teryx ski jacket, pointed out young guillemot birds—relatives of the puffin—floating in the water. They'd just left the nest. They were too fat to fly, so they had to just bob there for a few more hours or days until they'd lost weight—as easy prey for hunters as might be, for instance, a fledgling country for foreign oil and mineral concerns. But Minik thought the birds were hilarious. Every time we weaved around one, he pointed it out, then giggled like a madman as it vainly flapped its wings.

My LODGING IN UPERNAVIK was a yellow two-story house just off the paved street. I'd found it by a mailing a villager who told me to call another villager, who'd sent a silent Inuit woman to meet my flight. She put my bag in a taxi, drove me to the house, wrote down the number of kroner I owed her (450, then about $90), and handed me a key and left. A few hours later the door opened again, and the woman ushered in my surprise roommates: a young Dutchman and an older Dane, both scientists with GEUS, the Danish geological survey.

The scientists had come to retrieve an instrument left on the ice cap at seventy-six degrees north—some two hundred miles and two hours of flying time away. It was a ten-foot-tall metal tripod with a hard drive, a solar panel, and various sensors meant to track glacial melt. It had stopped working, but this patch of the otherwise shrinking ice sheet wasn't moving particularly quickly. GEUS was nevertheless spending tens of thousands of dollars to get it back, and because they had extra seats, I, too, was given a chance to benefit from Denmark's largesse. Helicopter charters in Greenland were otherwise expensive and near impossible to find: Oil and mining firms had booked them all up.

The helicopter was a single-rotor Bell 212, immaculately red like the rest of the Air Greenland fleet. One morning, just after dawn, we climbed in, and it lifted us above the town, above the fjords. Out the window were fog banks and empty islands, then a single iceberg in a windswept bay, then hundreds of icebergs, then thousands. The pilot, a Norwegian, flew between them, yards above the water. Then we climbed again and followed the ice cap north. Where glaciers were calving, spilling into the ocean, the seawater had frozen over during the night. On the ice cap itself, the surface was heavily crevassed and endless, a pattern of thousands of parallel cuts. I saw blues and grays and whites and browns, the red of the rocks, the orange of the rising sun. What I didn't see were people: On various islands were remnants of villages, stone walls and abandoned buildings, but the landscape had been further emptied as hunters moved to the cities.

I shared my window with a Dane who'd also heard about the helicopter's free seats. His name was Nikolaj, and he was a lab tech at the Upernavik hospital. He and the pilot also co-owned a kayaking business that rented out boats, dry bags, satellite phones, and polar bear protection in the form of .30-06 rifles. Business was good. That summer, fifteen foreigners had come, including two Israelis who camped out on an island for a month.

We stopped for a mandatory refuel in the village of Kullorsuaq—the only sign of life was the howling of sled dogs—and I quizzed Nikolaj about the hospital. The doctors were all foreigners, he said. "They come for one month at a time. Obstetricians, maybe one week. It's like a vacation for them." I asked what he thought about the referendum. "People here are spoiled," he said. "They have no idea how much things really cost." Greenland should stick with Denmark but find a way to keep the oil money, the pilot suggested. I wondered aloud if Denmark was really so enlightened as to give up all that oil. "That just tells you something about the Danish people," said one of the GEUS guys. We spent ninety minutes on the ice cap, just long enough for the tripod to be taken apart and stuffed into a wooden crate, then flew back to Upernavik just in time for me to make the next leg of the road show.

In Uummannaq, a thirteen-hundred-person island town that was famous as the home of Siissisoq, a metal band that sang in Greenlandic about the slaughter of African mammals, the road show was joined by Greenland's then premier, Hans Enoksen, a fierce secessionist and unlikely mentor to Minik. He was a former town grocer who rose to power in 2002 after serving as hunting and fisheries minister. In a high school auditorium, I watched him take part in a four-on-one smack down of the only anti-secession politician. The school's main hall was bright and angular and modern, with vaulted ceilings and walls of art— triptychs of icebergs, a painting of bananas and grapes. Enoksen was stern and relentless, slowly pumping his fist in the air as he spoke. It was a full house, eighty or more citizens. "We have been a colonized people for three hundred years," he rumbled. "Now that we have this opportunity, how can we say no?"

The premier hired a blue powerboat the next day, and we headed off to visit nearby settlements. We pulled out of Uummannaq's harbor, past its helipad and heart-shaped, thirty-eight-hundred-foot landmark mountain, and into a broad channel between sheer cliffs of stratified granite. After a while the premier turned to me. "The American ambassador in Copenhagen has been very supportive of self-governance," he said, Minik translating. "Much more than any before him." (In a leaked cable, the ambassador, James P. Cain, bragged of introducing Enoksen and a future premier, Aleqa Hammond, "to some of our top U.S. financial institutions in New York.")

America's support was unsurprising. In 1946, Washington was so impressed with Greenland's strategic potential that America secretly tried to buy the island from Denmark for $100 million. The U.S. military still

ran Thule Air Base, a cold-war-era installation in Greenland's far
north that had more recently been used for the CIA's extraordinary-
rendition flights. Before that, we'd apparently lost a nuclear warhead
there—leaving local hunters eating radioactive fish and seals. Now that
we'd learned Greenland had a lot of oil, our companies were buying
exploration blocks.

I wondered if giving up Denmark meant embracing America. Not necessarily America as overlord—indeed, China would soon make a play for
Greenland's minerals—but America as capitalist ideal. Americanism—
the free market, the need for growth, the never-ending quest for oil. I put
the question to the premier: Was replacing Denmark's money with other
foreigners' money really independence? He didn't quite answer. "If oil is
discovered, foreigners will come no matter what," he said. "But after we
vote yes, they will be working for us." He pounded his fist against his
chest three times, then raised it to the sky. "This is what will change
under me," he said.

MONTHS EARLIER, I had attended the first annual Greenland Sustainable Mineral and Petroleum Development Conference, held in a Radis-
son in Copenhagen in May 2008. Only one native Greenlander gave a
talk, and he was almost indistinguishable from the other attendees,
nearly all of them middle-aged men, nearly all of them in blue or white
dress shirts. He presented the Greenland Secretariat of Ice and Water's
market research from Los Angeles and Tokyo. It was extremely promising. Bottled-water buyers knew next to nothing about his island, he said,
but they knew all they needed to know. "Their knowledge of Greenland
is limited to 'ice' and 'cold,'" he explained.

The other speakers, Canadians and Australians and Brits and Swedes,
veteran operators in Rajasthan and Guinea and Mongolia and the
Philippines, described the mineral rush: West Greenland gold discover-
ies and South Greenland gold mining; two-and-a-half-carat diamonds

found by the Canadian firm Hudson Resources; rubies drilled by the Canadian firm True North Gems; open-pit molybdenum mines proposed by the Canadian firm Quadra Mining; and uranium and rare earth minerals finds by an Australian-owned, eventually Chinese-backed company with a local name, Greenland Minerals and Energy. A representative of GEUS spelled out Greenland's petroleum prospects. The miners discussed the island's tough logistics but "world-class commercial terms." If you could get there, they implied, these Inuit would let you drill anywhere. The prospect of having mining underwrite their independence from Denmark had made Greenlanders very agreeable.

The speaker who most explicitly linked his mine's fortunes to climate change was Nick Hall, CEO of the British company Angus and Ross, who showed the room a photograph of a mountain of marble above a giant fjord: the Black Angel. The zinc deposit here is one of the richest on the planet. It was discovered in the 1930s, explored in the 1960s, and mined between 1973 and 1990 via tunnels dug high above the fjord. Then it was abandoned. Hall's company took over the lease in 2003, when zinc prices were about to rise, and in 2006 two geologists on a day hike discovered a deposit as pure as the original at the edge of the retreating South Lakes Glacier. Before the melt, it had been hidden by a hundred-foot-thick wall of glacial ice. Along with the extended shipping season, it was, Hall had admitted, the "upside of global warming."

After he finished, Hall was surrounded by financiers: Australians in pressed suits representing British money, handing him business cards. A Canadian approached to offer the services of her logistics company: nurses, medics, and other camp staff. I approached him, too, and asked if I could visit the mine when in Greenland. Angus and Ross, by all accounts responsible and well-meaning, represented the uncomfortable reality faced by the northern "winners" of global warming, be they Inupiat, Greenlandic, Icelandic, or Canadian: Local residents did not have the capital, expertise, or population base to transform the Arctic on their own. There was the danger that they would get most of the change and most of the degradation, and wealthier outsiders would get most of the profits.

When I set out to reach the mine from Uummannaq, I sped out of the same harbor as I had with Premier Enoksen and into the same broad channel, but this time the boat captain was Danish, and he was working for the British. We left the channel and crossed a choppy stretch of open water, then hugged another set of cliffs. Entering a long fjord, we waved at fishermen and slowed down to watch a village woman butcher a seal on a rock. The fjord narrowed, and the water became glassy. After two hours, the namesake Angel rose before us: a Rorschach blot of ghostly black zinc two thousand feet up a mostly white cliff.

It was the end of the summer work season, and the mining camp was nearly empty. It was a series of prefab buildings on a man-made plateau, surrounded by the crumbling concrete and rusting machines of the original operation. Next to the harbor was a cable car—purchased secondhand from the Swiss ski area Disentis, in the melting Alps—that would someday span the mile-wide fjord to the mine. The buildings contained bunk rooms and a lounge with comfortable couches, a wide-screen TV, and a fast Wi-Fi connection. Inside the lounge, Tim Daffern, the mine's Australian operations manager, told me his company's game plan.

After pulling out the two tons of zinc left in the original mine—the support pillars, mainly, which they'd replace with cement pillars—they would focus on the deposit at South Lakes Glacier. It was certain to keep retreating: They'd commissioned a study by GEUS and some British scientists to be extra sure. If the original mine would last five years, South Lakes would buy them another decade. A third deposit they had identified could buy two more years, a fourth, three more—glaciers shrinking all the while. "Anywhere the ice retreats," Daffern said, "we'll explore."

Daffern's predecessors had dumped their tailings in the fjord. The waste was 0.2 percent lead and 1 percent zinc, and it had rusted before it could sink into the anoxic depths. Every spring, a rush of melting water had spread the waste farther. It had been ingested by blue mussels, and fish had eaten the mussels, and seals had eaten the fish, and on it had gone up the food chain. After seventeen years of mining, it had taken another seventeen years for the fjord to recover.

Daffern and I took a walk in the rain, climbing above the mining camp until we had views of the entire fjord, the fog banks, the seracs of the Alfred Wegener Glacier. I ventured into an old mine shaft until its slope steepened and became a sheet of ice. Daffern pointed out another shaft, where they'd found bags of chemicals that had been dumped and then sealed in by a bulldozer sometime in the 1980s. Daffern promised to do things differently. He also promised, just as everyone had at the mining conference, to hire as many locals as possible. When we returned, we ate an incredible, five-course lunch prepared by the camp cook, a guy named Johnny, who was Filipino.

ON DAY SEVEN of Minik's self-governance tour, after seven meetings in seven villages and towns, the politicians relaxed in a government guesthouse outside the Quarsut Airport, waiting to go home. Our flight wasn't until 4:30 p.m., and we had the entire day off. There was a buffet with muesli, yogurt, and fresh-baked bread. The TV was on; we pulled out cell phones and laptops and flipped through the newspaper. Then the premier walked in and announced that a hunter's boat was ready to take us on a quick visit to the village of Niaqornat, population sixty-eight, more than an hour up the Nuussuaq Peninsula. Going out again was masochism. Only Minik and I agreed to join him.

The open boat was maybe fifteen feet long. We hopped in at a gravelly beach below the airstrip, timing the surf so our feet didn't get wet. Minik put his laptop in a plastic bag. He and I kept low out of the biting wind, but the premier, wearing jeans, thin gloves, and a baseball cap, stood in the back of the boat, watching the coastline zip by.

The water was smooth, and there were beaches the whole way; above them, slopes rose steeply to six-thousand-foot summits already covered in snow. We passed seals and house-size icebergs and finally looped into Niaqornat's natural harbor. The village was stunning, on a spit of low-lying land between an ocean-side turret of rock and the white peaks of the peninsula. There were bright wooden houses but no cars. There were racks where villagers were drying junk fish for the sled dogs and strips of halibut and seal for themselves. Open boats and icebergs shared the harbor. The sun was shining. It was, for once, the Greenland of my imagination—and perhaps that of the premier's as well.

The meeting was held in the schoolhouse, and a quarter of Niaqornat showed up, if you count the baby. To make a projector screen, they flipped around a big map of Greenland and hung it over the blackboard. Above the map were diagrams so the children could learn the Danish names of everyday items: balloon, anorak, king, cigarette. As the premier talked, I checked out a poster showing eight local whale species and their key specs: weight, top speed, length, amount of time they could hold their breath. A man in a T-shirt that read "Deep Sea Shark Fishing" asked about money, and Minik flipped through some slides I hadn't seen before: projections of mineral revenues skyrocketing into the future. One showed the oil blocks that Greenland had already sold off to foreign firms. They were just on the other side of the peninsula.

We had lunch in the home of one of the premier's supporters, a great hunter whose walls were decorated with narwhal tusks and walrus skulls and pictures of dead polar bears. He laid out dried, jerky-like whale meat, then served us cold narwhal skin, which his daughters and the premier sliced into chewable chunks. His CD collection and computer were in the corner, along with his daughter's pet gerbil. His teenage son walked in with a premade sandwich and stuck it in the microwave. The premier gorged on narwhal. "If we did not eat what the sea gives us," he said, "we would not be here." When we reached the dock to meet our boat, the village had gathered to see us off, and someone had distributed little Greenlandic flags, which the citizens waved back and forth until we were out of view.

A few months later, Niaqornat would become one of a handful of villages to vote 100 percent in favor of self-governance. The referendum would pass by 75.5 percent across Greenland, but in tiny Niaqornat there were no doubters.

EARLY IN OUR TOUR, Minik had worried aloud that he was forgetting much of the philosophy he'd studied. "I've been too much into politics," he'd told me. But during our last conversation, he became a philosopher again, pondering not just the morality of secession but the means to this end. We were in Ilulissat, Greenland's big tourist town, where we had a final layover. Nearby was the fastest-sliding glacier in the Northern Hemisphere, Sermeq Kujalleq, which spits thirty-five trillion liters of ice into Disko Bay every year.

I had spent the early evening on the boardwalk of the Hotel Arctic, a cliff-side landmark that happened to be hosting the Nordic Council's Common Concern for the Arctic conference: European dignitaries wearing somber colors and fretting abstractly about the warming north. Peering into a bay full of icebergs at sunset, I heard one of them chat up an attractive blonde by rattling off facts about the coming doomsday. His tone was solemn, his voice almost a whisper. "I don't mean to scare you," he murmured. It was the first time I'd heard someone try to use climate change to get someone else into bed. "I really don't mean to scare you," he said again. She didn't look scared at all.

Upstairs, Minik and I ordered hamburgers at the bar and stared out at the lights of Ilulissat. "It's so strange," Minik said. "The more the ice cap melts, the more Greenland will rise. These other countries are sinking, and Greenland is rising. It is literally rising." Below us, the dignitaries filed into their banquet. "We know Black Angel was really bad for the environment the first time," Minik continued. "It ruined the fjord. Is it okay to ruin three or four fjords in order to build the country? I hate to even think this, but we have a lot of fjords. I don't know. That'd be utilitarian philosophy, wouldn't it?"

He shook his head. "We're very aware that we'll cause more climate change by drilling for oil," he said. "But should we not? Should we not when it can buy us our independence?"

FOUR

FATHER OF INVENTION

ISRAEL SAVES THE MELTING ALPS

The winter after Greenland voted yes, I traveled to where the melt was entirely less welcome. The Pitztal, or Pitz Valley, is thirty miles west of Innsbruck, the capital of the Austrian state of Tyrol. To reach it, I drove a rented Ford Fiesta at car-rattling speeds down the autobahn, then veered south at the village of Arzl, which had a host of small hotels and a church with an onion-domed steeple. I followed a two-lane road uphill through more postcard villages, passing fields, cows, and herders' huts—remnants of the pre-tourism economy—as the wooded walls of the valley steepened and busloads of Dutch vacationers appeared. After half an hour, the valley seemed to come to a head. There was a parking lot, a ticket booth, and a tunnel bored into a mountainside—an underground funicular railway. I boarded it and eight minutes later was thirty-six hundred feet higher, staring at the Alps' most famous disappearing glacier.

One measure of the Pitztal Glacier's decline is that one of the ski lifts built atop it has had to be moved three times in twenty-five years. Another is the giant, insulated blanket the resort cloaks over the glacier every summer, hoping to slow the melt. As a whole, Europe's Alps have lost half their ice over the last century, one-fifth of it since the 1980s. The 925 named glaciers in Austria are receding at an average of thirty to fifty feet a year, twice the rate recorded a decade ago. What has brought international fame to the Pitztal in particular—reports on NBC, articles in *National Geographic* and *USA Today*—is less the rate of its melt than the last-ditch absurdity of its glacier blanket. Workers cover nearly thirty acres at an annual cost of $120,000, preserving five vertical feet of snow per season. The technique has spread to Germany's Zugspitze and Switzerland's Andermatt and Verbier. But it only partly works: Covered or not, the Pitztal Glacier has already shrunk so much that it now peters out seven hundred feet above the lift station. During the all-important shoulder seasons—Pitztal is the highest of the five Austrian resorts used for fall and spring skiing—the last section of the ski run is a pile of jagged boulders.

Some 80 million tourists come to the Alps each ski season. Some 1.2 million Tyroleans, including nearly everyone in the Pitztal, depend on glacier skiing for their livelihoods. But across Europe, across the world, an economy is imperiled. In early 2007, slopes were bare the week before the famed Hahnenkamm World Cup race in Pitztal's neighboring Kitzbühel, and helicopters had to fly in 160,000 cubic feet of snow at a cost of more than $400,000. That same year, a British investor bought Switzerland's low-lying Ernen ski area for 1 Swiss franc; resort managers in Whistler, Canada, began using computerized global-warming simulations to choose the site of their next lift (answer: try higher uphill); Bolivian scientists declared that the country's lone ski area, 17,388-foot Chacaltaya, would lose its glacier entirely within three years (they would be proven right); and the Australian-designed, indoor, revolving Ski-Trac was loudly promoted "as the answer to the problem of climate change." The next winter, dome-encased indoor ski areas, including the seven-hundred-vertical-foot SnowWorld Landgraaf in the low-lying Netherlands, were officially added to the European race circuit.

Snowmaking has become a billion-dollar global industry. Cannons now spray man-made snow on nearly half of Austria's ski terrain, sucking up roughly 500,000 gallons of water per acre of artificial snow. Across the Alps, snowmakers use more water than does Vienna, a city of 1.7 million people—as much water per acre, it turns out, as a typical field of wheat. But traditional snowmaking, no matter how much it drains Europe's ponds and lakes, cannot secure the Alpine economy. It requires perfect conditions—below-freezing temperatures, a humidity of less than 70 percent, and minimal winds—and at Pitztal at least, these conditions are rarely present anymore when they're most needed.

When I visited the resort, the mountains were blindingly white, drenched in full sun and cloaked in natural February snow. The boulders were buried, while parts of the insulated blanket remained visible, its ridged individual sections poking out of the slope like vertebrae. From the end of the funicular at 9,318 feet, I followed the crowds to ride a clear-walled cable car to 11,286 feet, looking out at a vast fishbowl of a basin split by jagged ridges. A blast of cold wind met me at the top, and I quickly clicked into my skis. I dropped through patches of ice to an expanse of soft powder that led to a groomed run that led, eventually, to a modernistic, slate-paneled, fifty-foot-high cement building—the reason I'd come. The building housed one of the world's first models of the IDE All Weather Snowmaker, a $2 million device capable of shooting out thirty-five thousand cubic feet of snow in twenty-four hours at any temperature on any day of the year. For Pitztal, it was the latest salvo in the war against melt. For me, after travels in Alaska, Norway, and Greenland, it was a symbol of a new kind of climate response. Here, as in many of the places I would soon visit, the effects of global warming were no boon. They were a problem. The upside, if any, was in selling the best Band-Aid.

A ruddy-faced lift manager named Reinhold let me into the building, and he stood with me as I stared at a giant white cylinder, a welter of tubes and pipes, and a row of gray instrument panels lining the back wall. Neither of us could read the labels. They were written in Hebrew.

"THE ECONOMIC IMPACT of Global Warming Is Beginning to Show,"
the press release had read. "IDE's All Weather Snowmaker brings under
your control what previously could not be controlled!" It sounded like
snake oil, but the pitch that had attracted the Austrians carried a good
pedigree. It came from a nation with a history of overcoming the worst,
from a corporation—Israel Desalination Enterprises—already making
millions off climate change by wringing the salt out of salt water. The
reason I traveled to Austria, and soon to Israel, is that the small story of
the snowmaker—and its intertwining with desalination—represented
the perfection of a rosy ideal: that innovation and market forces, when
unleashed on climate change, can save us from it. Both Israel and IDE
also embodied a worldview that was at once empowering and dangerous: that solutions are worth their side effects. And their machines were
more proof that technological defenses against climate change are generally going first to people who can afford them, those who are emitting
the most carbon, who are taking care of themselves before turning to
the developing world.

How Israelis could know about snow was explained to me a week
after I left Pitztal, when I met IDE's technology chief and self-proclaimed
"best skier" at his home thirty minutes from Tel Aviv. Avraham Ophir
was dying of cancer, a white-haired man with a soft voice. His two col-
leagues who sat with me on a couch, Moshe Tessel and Rafi Stoffman,
looked at him with a mixture of fondness and awe. He was the institu-
tional knowledge of IDE, a now iconic Israeli company, and the hero of
one of its two gulag creation stories. He sat back in a red leather chair
and began telling it.

"Look, it's a long story, but I try to make it short," he said. "I was born
in eastern Poland, in a town called Bialystok. My father owned the factory that produced turpentine, which comes from the wood of trees in
this region. Now, in the beginning of the Second World War, we were

first occupied by the Germans for two weeks, and then the Russians came in. My father being a capitalist, he was taken prisoner and sent to a gulag in northern Siberia. And we, as the family of a prisoner, were sent to the south of Siberia, actually northern Kazakhstan." There, Avraham was forced to learn how to ski. "You would take two simple wooden planks that were very strong," he said, "and you would put a leather strip around it, and with your normal boot you would enter the leather. This is how we would go to school." Normally, an older boy led the students, because of the wolves. When blizzards, known as *buran*, came, they found their way by triangulating off telephone poles spaced every 150 feet near their route. They survived the long winters by eating fish caught during summers and cured with salt.

The story of the snowmaker also started in Siberia. In Russia, Avraham said, "there was a Jewish engineer by the name of Alexander Zarchin. This engineer was a Zionist. Being a Zionist and being a technologist, the Soviets sent him to one of the gulags, the same one as my father. And in Siberia it's very cold, but the summer did not have any rain. The gulag was close to the Arctic Ocean." The labor camp needed a source of drinking water. So in the summer, Avraham said, "they would open a gate and let seawater enter a lagoon. At the end of summer, they would close the gate, and the upper layer of the lagoon would freeze." When it did, the salt and water were forced apart. "By nature, ice crystals from seawater are pure water," he explained. When summer came again, the surface began to melt, flushing any residual brine from the ice pack, and Zarchin and the other prisoners began pumping liquid from the saline depths of the lagoon. They measured its salt content as they pumped, and once it was low enough, Avraham said, "they closed the gates and let the sun melt the rest of the ice—and they had drinking water." He looked at us proudly. "So you see," he said, mangling the phrase I would soon hear everywhere in Israel, "need is the father of invention."

After the war, Avraham was allowed to return to Poland, then smuggled with a group of Jewish children, Holocaust survivors, across the Alps to Italy—and eventually to the newly declared state of Israel. Alexander Zarchin, his future boss, also fled from the gulags to Israel, where he soon found fame as an inventor. In 1956, the country's first prime minister, the water-obsessed David Ben-Gurion, gave Zarchin a quarter-million dollars to build a pilot desalination plant. In 1960, *Look* magazine declared that what was being called the Zarchin process could "have more significance than the atomic bomb." Zarchin's trick was to replicate the Siberian freeze using a vacuum chamber: when pressure drops below four millibars, chilled salt water becomes ice, and thus becomes desalinated. His project, eventually incorporated as the for-profit IDE, became a vehicle for both capitalism and nationalism. "He wrote down the patent when he saw that the country needed water," said Avraham. "Most of Israel was a desert at that time, but in the Bible the country was full of trees. You read that the son of David, Absalom, he was running away on a horse, and his hair got caught by the branch of a tree, and this is how he was killed." Avraham gestured out a window toward his lush garden. "We decided we were going to make this country look like it did before," he said. "And the people who came from Eastern Europe and other places, they wanted to convert it into something that reminded them of where they had lived before."

When foreign investors came, a nervous Zarchin covered his machines' dials with cloths, determined that no one should steal their secrets, or his profits. But vacuum desalination was quickly supplanted by more efficient reverse-osmosis techniques, and it took IDE forty years to find a real use for the Zarchin process. The eureka moment—this one belonging to Avraham—came in South Africa, where an IDE vacuum-ice machine was commissioned to help cool the world's deepest gold mine, two miles below the surface of the Earth, where workers faced 130-degree temperatures.

It was back in 2005, Moshe explained. He and Avraham were in South Africa on a site visit, testing out the mine's newest machine.

Avraham saw a pile of snow produced in the heat of the African sun, and his eyes lit up. "Moshe, get me some skis," he commanded. Moshe went into Johannesburg and found some skis. "At lunch, he had a big exhibition," Moshe says. "I told him, 'Avraham, I'm impressed you are a good skier for your age'"—he was seventy-two years old—"'but before we take it to the Alps, let's find a specialist.' I looked for one on the Internets."

The specialist Moshe found, a Finnish Olympic coach, was flown down to South Africa. According to Moshe, he declared the pile "fine snow for ski—not powder, like in Aspen, but it's what the professionals call spring snow." IDE then flew a dozen ski-area executives down, Moshe says, "and we built two snow mountains, and we spent two days with the guys, and we ate, we drink, and already after this I get two orders." Iconic Zermatt, the village below the Matterhorn and the now melting, shifting Swiss-Italian border, got the first IDE snowmaker. Pitztal got the second.

In 2009–2010, the first full season it deployed its new snowmaker, Pitztal became the first ski area in the Northern Hemisphere to open. The date: September 12. After the debacle that was the 2010 Vancouver Olympics, when helicopters had to ferry in snow for the events on the barren slopes of Cypress Mountain, Russia—the host of the 2014 Winter Olympics—asked IDE for a demo at Pitztal. Officials were impressed. Russia began stockpiling snow underground and under tarps, with plans to have as many as three thousand tons stored when the games began.

"We managed to sell snow to the Eskimo," Avraham said.

"Now I want to sell sand to the Bedouins," said Moshe.

"They have no money," Rafi said with a laugh.

FOR IDE AND THE REST of the desalination industry, there was an aspect of the planet's ice loss that was even more auspicious. What comes after melt is drought. In the Alps, no less than in the Himalaya or Rockies or Rwenzoris or Andes, disappearing ice is disappearing water storage. Glaciers are reservoirs. Snowfields are reservoirs. In winter, they grow with precipitation, trapping it uphill. In summer, just when it is most needed, their water is slowly released. Shrinking glaciers imperil the water supplies of seventy-seven million people in the tropical Andes, along with the hydropower providing half the electricity in Bolivia, Ecuador, and Peru. In Asia, two billion people in five major river basins—the Ganges, Indus, Brahmaputra, Yangtze, and Yellow—depend on Himalayan meltwater. The range's glaciers, which irrigate millions of acres of rice and wheat in China, India, and Pakistan, lose an estimated four to twelve gigatons of ice a year. In Spain, which is becoming so dry so quickly that some warn of "Africanization"—of the Sahara jumping across the Strait of Gibraltar—the Pyrenees have lost nearly 90 percent of their glacial cover. A century ago, the glaciers feeding such agriculturally important rivers as the Cinca and the Ebro stretched 8,150 acres across the range. They now cover 960 acres. And even in the United States, millions of people depend on glaciers and winter snowfall: Southern California, kept green by mountain-fed rivers, especially the Colorado, is in danger of losing 40 percent of its water supply by the 2020s if melt in the Rockies and Sierra Nevada continues apace.

In a sense, Israelis understood better than anyone what it was like to descend into drought. They knew what to do. Coming here from Europe, as Avraham had explained, they had faced a changed environment—hotter, drier, and less hospitable than what they had known before—and they had faced it head-on. Zionism had been guided by Enlightenment ideals: faith in reason, faith in capitalism, faith that any problem, even the treatment of Jews in Europe, had a rational solution if man was rational enough to find it. The first Israelis did not bow before nature. The Enlightenment answer to water scarcity, then as now, was to seek the silver bullet—an engineered solution, a supply-side solution.

"For those who make the desert bloom there is room for hundreds, thousands, and even millions," Ben-Gurion had written in 1954, when he himself moved to the Negev Desert. Next, the prime minister underwrote Zarchin's test plant in the Negev. He began funding cloud-seeding operations, including 1960's Operation Rainfall, in which silver-iodide dispensers were attached to the wings of fighter jets. He built the National Water Carrier, eighty miles of pipes, canals, tunnels, and reservoirs, to move water from the Sea of Galilee, in the relatively wet north, to the Negev, in the bleak, underpopulated south. Some of the fixes failed. Some had side effects. The water carrier would stoke war with Syria over the headwaters of the Jordan River, and it would soon feed a massive, export-focused agricultural industry. To export a gram of wheat was the equivalent of exporting a liter of water, so eventually Israel would export the equivalent of 100 billion liters a year. But at the time, few people questioned if any of this made sense.

We were all becoming Israelis now. In Peru in 2009, a scientist won a World Bank award for his proposal to paint the Andes white and repel the sun's lethal heat. In India's Ladakh region, a retired engineer built a $50,000 artificial glacier in the shadow of the Himalaya, collecting runoff in rock-lined ponds that would freeze and attach to an existing glacier in winter. In Spain, Barcelona became the first city in mainland Europe to resort to emergency water imports: five million gallons transported in 2008 in a converted oil tanker. In China, the central government prepared to divert rivers at a scale the world had never seen: The $62 billion, three-canal, 1,812-combined-mile South-North Water Transfer Project will someday move 4.5 trillion gallons each year from the Tibetan Plateau, home to nearly forty thousand melting glaciers, to the cities in the country's arid, industrializing north. More than 300,000 citizens were being displaced to make room for canals and pipes. While China waited for Tibet's water, its Weather Modification Office—thirty-two thousand on-call peasants manning thirty bases across the country at a cost of

$60 million a year—was shelling its skies with rocket launchers and 37-millimeter anti-aircraft guns, delivering silver-iodide pellets in hopes of inducing rain. And in China, India, Peru, Spain, and seemingly every country where rising heat and melt had induced drought, massive desalination plants were also on the rise. Between 2003 and 2008, 2,698 plants were built worldwide, and hundreds more were under construction.

By the time I visited Israel, IDE was responsible for nearly four hundred of the world's desalination plants, including what was then the biggest, most efficient, and most celebrated: the 86-million-gallon-per-day (mgd) plant in Ashkelon, Israel, next to the Gaza Strip at the edge of the Negev. IDE's partner at the plant was Veolia, the world's biggest water company and one of Deutsche Bank's top stock picks. After Ashkelon, IDE had won contracts to construct the largest plant in China, a $119 million job; a 43-mgd plant in desiccating Australia, a $145 million job; and a giant, 109-mgd plant north of Tel Aviv in Hadera, a $495 million job. IDE was also part of the consortium building two contentious 50-mgd plants in Carlsbad and Huntington Beach, California. An engineer at the company leading the construction, Poseidon Resources, told me they would be able to create water with the exact mineral content and taste of Pellegrino. "People will drink Pellegrino out of the tap," he said, "and they'll take showers in Pellegrino."

Ashkelon met almost 6 percent of Israel's total water demand, a first step in the country's plan to get a quarter of its water from the sea by 2020. After subsidies, its price per cubic meter was just sixty cents—on par with tap-water costs in the United States, far cheaper than in parts of Europe. Once bought by the government, nationalized, and dumped into the National Water Carrier, its water was indistinguishable from the rest. But, as the Shell scenarios team highlighted in its exploration of the water-energy-food nexus, there was a problem: Desalination plants, even Ashkelon, use vast amounts of power. Power plants—whether nuclear, coal, gas, or hydroelectric—use vast amounts of water for cooling. If they are fueled by coal, or, to a lesser degree, natural gas, they also emit vast amounts of carbon. Carbon furthers warming, warming furthers drought, and desalination begins to resemble a snake eating its own tail.

If a glass of water is eight ounces, a glass of water from Ashkelon takes 10,200 joules. The plant runs on natural gas, making it cleaner than most, but that glass of water still translates to 0.0011 pounds of CO_2 emissions. If the average Israeli were to take all his or her water from Ashkelon, it would cause 0.6 metric tons of annual emissions—about half of what each person on the planet can emit if we're to someday halt warming. (Israelis currently emit about 10 metric tons, Americans 20 metric tons—both already well over that limit.) In California, the numbers are much worse. The Carlsbad desalination plant, the largest in the United States, may get most of its energy from coal. It will then be responsible for more carbon emissions—97,000 metric tons a year—than a dozen island nations. But no one claims desalination can save the world. Nor can any snowmaker save all the world's glaciers. They can only save the rich parts from the fate befalling the rest.

"IF YOU HEAR A WHISTLE, get under the car," the IDE engineer Elisha Arad said when he, Rafi, and I set out for Ashkelon. The Palestinians had just begun firing their first Grads, 170-millimeter rockets with enough range—8.6 miles—to reach Ashkelon from the Gaza Strip. The day before, a school had been hit, and a pair of tractors was needed to dislodge the rocket from its crater.

"We give them water; they give us rockets," Rafi complained. This was true but incomplete: In the previous month's offensive against Hamas in the Gaza Strip, which the Israel Defense Forces called Operation Cast Lead, Israel had damaged eleven wells, twenty miles of water networks, and more than six thousand rooftop water tanks. It then kept the borders locked down, which had the effect of keeping pumps, pipes, and cement out of Gaza—and making repairs impossible. A year after the operation, ten thousand people would still be without access to the water network, and Gaza's principal aquifer would turn saline. Palestinians would begin tapping into Israeli pipelines, and Israel would crack down on what was termed water theft. In Israel, per capita water consumption was 280 liters a day; in Gaza, it was 91 liters—below the 100–150 liters the World Health Organization says is necessary. Elisha suggested that Hamas hadn't hit the Ashkelon plant because it didn't want to: The plant was producing the water everyone needed. Rafi suggested it was because members of Hamas were bad shots.

Our SUV rumbled down an empty freeway into the Negev, past a plantation of cacti and a strange outpost of ten-year-old prefab homes in the sand, populated by Jewish settlers who'd relocated there from Gaza in a political tussle my hosts remembered little about. Elisha, a sixty-year-old who had a bald head and, when asked a question, a single, deep furrow in his brow, told me about the Dead Sea: It was supposed to be replenished by the Jordan River, but the National Water Carrier kept carrying water elsewhere. "Now it's dying, literally dying," he said. "It's losing one meter a year. We will face no water in twenty years unless supply is given." He claimed no one bothered to conserve: "Agriculture pays ten times less for water. If he pays ten cents, while I pay a dollar? Of course then he doesn't care."

Around a corner, just before reaching the desalination plant, we caught sight of a blimp hovering over the border with Gaza: a sentry system for rocket fire. "It gives the alert in the townships," Elisha explained. "It gives people time. Time, let's say, to find a shelter." Silent, all seeing, futuristic, the blimp must have looked oppressive to those on the other side of Israel's walls. But on this side it was comforting. The country's approach to drought—so many carbon emissions from so much desalination, a detriment to the world, in exchange for endless water, a boon for Israel—was similar, I thought. What made sense within the borders of one nation, especially one surrounded by enemies, would not always make sense outside it.

The desalination plant shared a site with an eleven-hundred-megawatt, coal-fired power plant, the largest in Israel. It was a location chosen not for cheap electricity but for the chance to minimize environmental impacts: The coal plant discharged hot water; the desalination plant discharged hypersaline brine. When the streams were mixed, each was diluted. "Now the fish are suffering . . . but, uh, less," Moshe had told me. The seventeen-acre facility was sprawling, eerily empty, staffed by just forty employees who took turns working eight-hour shifts, almost alone in an automated future.

Under a low-slung sky of cumulus clouds, we put on yellow hard hats. The Mediterranean was dark green, but the buildings were painted ocean blue. Elisha led us from one to another, from the settling ponds to the pretreatment pumps to the carbon filters to the synthetic filters. The plant used reverse osmosis, a rival process to Zarchin's first developed in the 1950s and 1960s by a Jewish chemical engineer in Southern California, then perfected in the 1970s at Ben-Gurion University of the Negev. Its main hangar was dominated by a quadruple array of reverse-osmosis membranes—forty thousand of them, arranged end to end in eight-inch pressure tubes. Everywhere we walked, I heard a rush of water. Outside, concrete bomb shelters were spaced every few hundred yards. "If you are within zero to 4.5 kilometers of Gaza, you have fifteen seconds," Elisha said. "If you are within 4.5 to 10 kilometers, you have thirty seconds." We were within zero to 4.5 kilometers. We stopped at a control room, and there, sticking out from a blue holding tank, a faucet appeared. Elisha produced a cup. The water tasted pure, perfectly natural, and after we all downed a first round, I had a second.

BACK IN TEL AVIV, I had a series of meetings with other water entrepreneurs, each hawking his own techno-fix. Israel was, to borrow a phrase, the "start-up nation," but this was the embodiment of another trend: Export-ready water technology seems to emit most readily from

water-strapped countries—Israel, Singapore, Spain, and Australia—
whose backs are up against the climatic wall. In Israel, traditional cloud
seeders were still legion, but one group of researchers also proposed covering a two-thousand-acre expanse of the Negev with black heat-
absorbing material to create an artificial heat island, thus inducing
downwind rains. In a high-rise downtown, I sat with executives at a firm
called Whitewater whose founder had once helped Domino's Pizza pen-
etrate the Israeli market. It had close ties to Prime Minister Benjamin
Netanyahu, and its approach was distinctly Israeli: It helped secure
nations' water supplies from contaminations and terrorist attacks.

My final meeting was with Dr. Etan Bar, a wiry, obsessed
environmental-engineering professor from Ben-Gurion University. He
had just patented a new way to battle drought, one so seemingly
revolutionary—and profit oriented—that I wondered if he could become
the next Alexander Zarchin. On what would be the rainiest day of the
year, we met in the lobby of the Tel Aviv Sheraton along with two of his
marketers, Yitzhak Gershonowitz and Levi Wiener. "You know," said
the latter, almost immediately, "necessity is the mother of invention."

Dr. Bar sat back and excitedly detailed his creation. "To take humidity in the air and turn it into liquid water," he said, "this process is
known even from the Bible times." At the beginning of winter, Jewish
people would say the prayer for rain, Tefilat Geshem. But at the beginning of the dry season, at the start of Passover, they would say the prayer
for dew—Tefilat Tal—for dew was considered a gift from God. "In the
southern part of Israel," Bar said, "you can see ancient fields that were
all irrigated based on dew, condensation of the air's humidity.

"You can believe or not that the globe is influenced by carbon dioxide," he continued, "but the bottom line is that it is being more and more
hot, which means two things: There is more evaporation from the sea,
because the water temperature is higher, and humidity on Earth cannot
be condensed because the air temperature is higher." In tropical countries that once had "rain almost every day, 365 days a year," the rainy

season was no longer so rainy. "You're walking around in a smog," he said, "because the humidity is there, but the rain is not. Everybody knows these facts."

Dr. Bar had designed a box that he said solved the problem, mimicking an ancient God: It sucked in air and spit out water. He outlined the steps: First, air is drawn across a desiccant that absorbs water vapor but not pollutants. Next, the desiccant is heated up, releasing the water into a vessel containing a much smaller volume of air. Finally, the heat is pulled out, to be used again in the process, and the vapor cools and condenses. "That's all," he said. "It's very simple. It's really simple. It's a filter that filters water from air. That's all." He claimed he only needed investors.

"Levi said that necessity is the mother of invention," he continued, "but I . . ."

"You don't believe so?" asked Gershonowitz.

"No, I don't believe so," Dr. Bar said. "I think that market needs are the mother of invention."

"If there's a market for anything," he said, "it's water. Nature is working for us. Nature is our best PR. Why? Because there is no water! Look at Cyprus. In Greece, it is the same. In Ivory Coast, no rain at all anymore. And I'm not talking about desert regions. I'm talking about places where they used to have a lot of water." Gershonowitz and Wiener nodded enthusiastically. "In 2020," Bar continued, "about one-third of the world's population will have no access to secured freshwater. The average water consumption globally is between fifty to a hundred liters a day. So now multiply that by 2.5 billion people! That's all you need. If you ask, what's a market potential, that's a market potential!"

PART TWO

THE DROUGHT

I swear by my life and my love of it that I will never
live for the sake of another man, nor ask another
man to live for mine.

—*John Galt*

FIVE

TOO BIG TO BURN

PUBLIC FIRES, PRIVATE FIREFIGHTERS

The first light we ran was at Main Street and Jamboree Road, near the Hyatt, and we ran it mostly because we could. Chief Sam flicked on his siren, and eighteen lanes of traffic froze in place. We nudged into the intersection. We accelerated. We swerved. We accelerated again. Our red Ford Expedition, topped with red lights, emblazoned with the word "FIRE," shot onto the 405, tires screeching. Car after car pulled over to let us by until, as we merged onto I-5, some idiot in a Civic didn't. "Look at this guy," Chief Sam muttered, and then he cut into the median to race past him.

The traffic died down near Disneyland, but the Santa Ana wind picked up. It was a hot, desert-born easterly that sucked any remaining moisture from the landscape. It funneled through the canyons in gusts, carrying brush, bits of cloth, plastic bags, and clouds of dust. The dust blasted across the freeway, ocean-bound, and our truck, now going seventy-five in the center lane, shook from side to side. Chief Sam, the head of a private army of for-profit firemen, stepped on the gas. He offered me a protein bar. He put headphones in his ears, picked up his BlackBerry, and began making calls.

A call to his crew: "Right now, Pump 31 should be partnered up or out patrolling. Pump 42 should be teamed up and ready to be deployed.

No delays. Just be out and about. A good staging location. Out of bed and get 'em married. Right now."

A live call to KTTV Fox 11 News. A call to one radio station. An interview with another ("There isn't even a fire season anymore. It's year-round"). A call to yet another: "Hi, I'm Fire Chief Sam DiGiovanna. I've done reports for you guys in the past. Do you want anything done this morning on these fires? DiGiovanna. *D-I. G-I. O. V. A. N-N. A.* And it's just very simple. Dee. Gee. Oh. Vanna . . ." We were still miles from the fire in Little Tujunga Canyon, only now passing downtown Los Angeles on our way north. Before some of his calls, Chief Sam turned on his siren. When he finished talking, he turned it off.

On the phone, he identified himself as the training chief at the Verdugo Fire Academy, his part-time job. He didn't mention the insurance companies he contracted for—least of all American International Group, or AIG, where he ran the firm's Wildfire Protection Unit. Insurance was the industry that had the most to fear from climate change—it was on the hook for increased hurricane and fire damages—but also, paradoxically, the most to gain: Its market was expanding, especially in places like the drought-racked American West. Insurers were funding climate research just as AIG's innovations in the financial sector were helping sink the global economy, and Chief Sam and I were driving toward the fire just as AIG was being bailed out by the federal government: $85 billion and counting.

We left I-5 for Highway 2. A white cloud of smoke was now visible in the distance, somewhere east of Pasadena, and Chief Sam switched on the news. The flames were burning right down to the 210 freeway, the announcer said. Wall Street was having a bit of a rally, up four hundred points. Chief Sam changed the channel to his favorite, Smooth Jazz 94.7, and an instrumental version of the Doobie Brothers' "Minute by Minute" filled the cab of the Expedition.

The first police line, a diagonal string of orange cones guarded by a single squad car, came after we turned onto the 210. Everyone was being funneled off the freeway, causing a traffic jam, but we accelerated in the left lane. The siren came on again. Chief Sam, a firefighter for twenty-nine of his forty-nine years, dark-haired and barrel-chested, looking official in his unofficial blue uniform and red fire truck, gave a convincing wave to the cop. The cop waved back. I watched the traffic pass in a blur until we crossed the cones and were completely alone. Suddenly everything had the look of war, the scent of smoke. Chief Sam showed no emotion, but he put down the BlackBerry.

The darkest clouds were chemical, from a fuel that was man-made, toxic—not chaparral, not wood. From a burning mobile home. A burning landfill. A burning panel truck parked on a ridge. Helicopters clattered overhead, dumping white clouds of water that seemed to have no effect. Gusts blew the vapor sideways, with the smoke. On the freeway, strike teams—five-vehicle convoys sent from neighboring cities—rolled by at eighty miles an hour. A semitruck from the Los Angeles County Fire Department passed us, towing a red bulldozer. Traffic signs appeared and disappeared, obscured by the haze. The hills themselves were turning black, and where the vegetation had burned away, the rocks were set free and small landslides littered the side roads.

The AIG team was waiting at a municipal park in Sylmar, the staging area for the fight against the five-thousand-acre blaze that was consuming Little Tujunga Canyon. Their trucks, Pumps 21 and 23, were red Ford F-550s with orange hoses and chrome panels—just two of the dozen such trucks commanded by Chief Sam. The men were in their twenties and thirties, clean-cut and bored. This neighborhood, though imperiled, wasn't quite rich enough to put them to work: AIG's Private Client Group insured and protected only homes worth at least $1 million. They were waiting for things to get worse.

THE LITTLE TUJUNGA fire was normal, Chief Sam had told me. This was October. This fit the pattern. He'd explained climate change's effect on firefighting the previous afternoon at the Hyatt, his favorite hotel, where he'd taken me after my flight landed in Orange County. We'd driven up in a rush, parked his truck in the emergency lane, and marched to the front desk, where he scored me a suite at the government rate. "Can we put the room on my Gold points?" he'd asked. "We'll, uh, both be staying there." He gave me a conspiratorial kick under the check-in counter. Then we sat in the lounge, snacking on complimentary wasabi peas. "I just love these things," he said.

The wet will become wetter. The dry will become drier. That which burns will burn more often. Whether California's drought at the time was specifically linked to climate change, scientists could not definitively say, but the state, along with much of the West, was on its way to becoming what the computer models predicted: a dust bowl, and too frequently on fire.

This fire had broken out after California's hottest spring in 88 years, its ninth-hottest summer in recorded history, its lowest rainfall in 114 years, its fourth month of government-declared drought, and its second, maybe third, year of unofficial drought. Arnold Schwarzenegger, the governor at the time, was showing up at rallies of Central Valley farmers, chanting in unison with them: "Ve need vater! Ve need vater! Ve need vater!" Homes would soon be lost to fire simply because the hydrants went dry.

"When I started in 1977," Chief Sam said, leaning forward in his chair, "there was a definitive season"—late summer and especially fall, after the hills had been baked dry, after the Santa Ana winds returned. But no more. In April 2008, he said, unusually high temperatures and low humidity fueled a 600-acre brush fire in the Sierra Madre. In 2007, a record twenty-one simultaneous Southern California wildfires had forced a record evacuation—346,000 homes. The second-largest fire in state history, the Zaca, had consumed 240,207 acres and $118 million in firefighting costs—another record. In May 2009, the 8,700-acre Jesusita fire in Santa Barbara would burn 80 homes and force at least fifteen thousand people to evacuate. In Southern California, the two fire years leading up to Little Tujunga were the worst two in the last two decades, blackening 1.3 million and 1 million acres, respectively. "With this global warming," Chief Sam said, "we're getting fires more often, in different areas."

Across the globe, the first decade of the new millennium was a decade of fire: Fire in Alaska and Spain and Siberia and Corsica and Bolivia and Indonesia and British Columbia. In New Mexico and Oregon and Colorado and Texas and Arizona. In the Black Hills of South Dakota and the swamplands of North Carolina. In Greece, the worst fires in half a century during the worst drought in millennia. In Australia, the worst fire in recorded history during the worst drought in recorded history. In Russia, fires so destructive that the president—Medvedev, not Putin—said out loud that climate change was real. The largest fires in Georgia's recorded history, in Florida's recorded history, and in Utah's recorded history. Across the United States, an average of seven million acres have burned each year of the new millennium—twice the 1990s average. Between 1986 and 2006, the number of major wildfires grew by 400 percent, the area burned by 600 percent.

The effects of climate change on wildfire were not limited to the lack of water or the heat of the hottest days. Early spring snowmelt meant longer growing seasons, eventually more fuel. Higher average temperatures meant summers were effectively longer, and fuels had more time to dry. Warmer winters meant parasitic larvae—pine beetles, spruce beetles, bark beetles, tent caterpillars—could flourish and expand their range, killing vast forests, creating more dead, desiccated fuel. If there is sustained drought, trees can't generate the chemicals to fend off the pests. In the western United States, spring-summer temperatures had risen just 0.87 degree Celsius since the mid-1970s, but the fire season was now seventy-eight days longer.

Los Angeles's meteoric growth, its climb into the wind-buffeted, fire-prone foothills of the Santa Monica and San Gabriel ranges, was also responsible. According to the state's forestry and fire agency, Cal Fire, 40 percent of California's 12 million homes were in areas of high or extreme danger. In Southern California alone, the Forest Service had identified 189,000 such homes constructed between 2003 and 2007. In the 1960s, wildfires burned 100 buildings in an average year. In the 1990s, they burned 300. The first decade of the new millennium, it was 1,500. "Normally, we consider the fuel to be trees and shrubs and brush," Chief Sam said. "But now it's not just trees. The homes are the fuel."

Already thousands of private contractors were battling wilderness forest fires on behalf of government agencies. Of the 280 pilots and ground-crew members in Cal Fire's aviation program, 130 were actually employed by DynCorp International. Of the Forest Service's $1.5 billion firefighting budget—about a third of its total budget—more than half ended up in the private sector, often with companies based in Oregon: the hub, for whatever reason, of the for-profit fire industry. The modern, privatized fire camp might have crews provided by Oregon's Grayback Forestry or GFP Enterprises, air support by Oregon's Precision Aviation, and catering by Washington State's OK'S Cascade Company. It will have mobile showers, mobile laundry, and mobile offices, and also air-conditioning, Internet connections, and tents with floors. Firefighting cost the federal government more than twice what it had cost a decade earlier.

If the first way to profit off climate change—that found in the Arctic—was to expand, to push into virgin lands and virgin resources, this was a new phase. The opportunity here was also a kind of growth, yes, but it was growth born of scarcity, of someone else's crisis—the zero-sum economics of distress. For there to be winners, there also necessarily had to be losers.

The key innovation by Chief Sam's company, Firebreak Spray Systems of Hood River, Oregon, had been to contract for the insurance industry, not the government. (Though in the case of AIG, the two would soon temporarily merge.) Founded by the entrepreneur Jim Aamodt, who invented the sprayers that keep produce fresh at the supermarket, and

Stan Brock, a former tackle for the New Orleans Saints, Firebreak had a proprietary system to coat houses with liquefied Phos-Chek—the same Monsanto-developed chemical retardant used by the Forest Service. The spray was colorless and harmless, Chief Sam told me, and it could protect your home for up to eight months, far longer than rival gels and foams.

In 2005, Firebreak went to work for AIG's insurance division, increasing the fleet of the division's Private Client Group from two to twelve trucks and expanding its reach from fourteen elite California zip codes—90049, 90077, 90210, and so on—to nearly two hundred, plus zips in Vail, Aspen, and Breckenridge, Colorado. Chief Sam joined the company in 2006, after five years as fire chief in Monrovia. He'd been planning a second career in executive coaching until he read about AIG's new wildfire unit in *Fortune*. "It was the thing of the future," he told me, "and I wanted to get in on the ground floor."

Firebreak was growing—Chief Sam and his friend George had just started a two-truck pilot program for Farmers Insurance—but suddenly there was competition. Chubb insurance protected policyholders in thirteen western states through Montana's Wildfire Defense Systems, which sprayed homes with the rival Thermo-Gel retardant. Fireman's Fund contracted San Diego's Fireprotec to clear a defensible space around clients' homes and offered evacuation services to its richest customers. San Diego's Fire-Pro USA sprayed homes with patented FireIce gel. Wildomar's Pacific Fire Guard deployed "the Navy SEALs of firefighters" to spray homes with GelTech retardant. Carmel Valley's Golden Valley Fire Suppression offered spray-foam services on Craigslist as well as "Land Clearing with use of a goat herd." It beckoned customers with an online survey. Question 6: "If you could have a private fire force that would specifically work to save YOUR home in the event of a threatening wildfire, and the price of this protection would be $35,000 (financing available), plus $1,600 per year thereafter, how likely would it be for you to hire them?"

Firebreak's trucks were outfitted with topflight communications systems, Chief Sam bragged, including RedZone mapping software that predicted a fire's course and revealed clients' addresses with a "tap on the dot." Unlike overstretched public brigades, Firebreak could afford to be better. "To be honest with you," he said, "we're probably more sophisticated than a lot of municipal agencies."

If private firefighting sounded like a libertarian dream—private industry stepping in where government falls short—in fact it was just that. As California burned, Adam B. Summers of the Reason Foundation, a free-market think tank historically funded by Shell, BP, Exxon-Mobil, and the climate-change-denying Koch brothers, contributed an op-ed to the *Los Angeles Times:* "Tap private-sector resources to improve fire protection." California was a high-tax, high-regulation state, he wrote, and this stifling business environment was driving jobs away. To increase taxes to fight more fires would only make it worse. "State and local governments can better provide fire protection services by tapping military resources and private-sector resources," he declared. "The private sector has a long and distinguished history of providing high-quality services such as paramedic services, security services, and, yes, even firefighting services at lower costs." Summers singled out Firebreak and AIG for praise. "As is the case in numerous other outsourced services," he wrote, "private fire protection contractors oftentimes provided equal or better services at significantly lower costs."

"We're in a tough economy," Chief Sam told me at the Hyatt. "It's important that local governments start working with privatization. Municipal agencies can't do it all on their own." We'd downed the wasabi peas. He summoned a waitress: "Hey, I hate to say this, sorry to bother you, but can I get some water from you?"

CHIEF SAM AND I LEFT the staging area and drove uphill, toward the fire, to check on the pilot program for Farmers Insurance. It was the new team's second day facing an actual blaze, and Chief Sam was still looking to build market share. He'd just directed Pump 43 to come here from San Diego, leaving the city exposed and worrying the people back at Farmers. We pulled over to check in. "I'm on the scene right now," Chief Sam reassured his Farmers contact as we stared at the walls of a freeway underpass.

We drove on to the Little Tujunga police line, where residents were amassed on the sidewalk, carrying their photo albums in pillowcases, their flat-screen TVs in cardboard boxes, their cell phones in their hands. We crossed the line with our lights flashing, nodding our heads at the police. Beyond it, garbage cans still sat curbside at every empty house—it must have been trash day—and the wind was toppling them, spilling their contents onto the street. A few stragglers in gas masks were defying the evacuation order. A kid rode his bicycle in circles in the middle of the road. An old man in a flannel shirt sprayed the sidewalk in front of his home with a garden hose.

Farmers Pump 25 was parked alone on a side street. George, an amiable man with gray hair and a gray mustache who'd fought alongside Chief Sam for decades before both retired from the Monrovia Fire Department, was in the driver's seat. He started up the engine, and both trucks charged uphill until we reached a modest, single-story home—Farmers was less exclusive than AIG—that was in little immediate danger. George's young partner hopped out wearing a yellow helmet and yellow protective gear, unspooled an orange hose, and tugged it up a set of brick steps. He squeezed the nozzle, coating the home's already sickly grass in Phos-Chek. Chief Sam encouraged me to take some photographs. I took some.

Then we waited, and I began to see that the ethical problems some people had with services like Firebreak—that they profited off disaster, that they protected only the wealthy—were secondary to another problem: Firebreak had a hard time protecting anyone at all. The list of homes we should spray—our "priority list"—came from dispatch up in

Oregon, which was supposed to determine which way the fire was going and which Farmers homes were in its path. But the fire was barely moving, because the Los Angeles Fire Department had it nearly contained. And the pilot program was so new, the dispatchers so unpracticed, that they seemed to struggle to find any Farmers addresses. We waited for orders outside the house we'd just sprayed, then on broad Gavina Avenue where it crossed Pacoima Wash, then uphill amid the neighborhood's newest, largest homes: palm trees, stucco roofs, territorial views, proximity to the flames. Forty minutes passed. A Skycrane helicopter and super-scooper airplane dumped retardant on the hills. Dozens of public firefighters, free to simply fight the fire, rushed past. None acknowledged us. Finally George got another address.

Chief Sam and I followed George's Pump 25 past a clump of "For Sale" signs shrouded by smoke, down one block, peering at house numbers, and down the next. We lurched forward, then braked, watching the pump's taillights flicker on and off. Chief Sam was getting agitated. "George, is that one of ours?" he asked over the radio. "That house on the corner, is it one of ours? Well, find ours. Is that one ours? Let's find ours and spray it."

Two hours after we arrived on the scene, I watched George spray a second Farmers property, a two-story stucco home in a subdivision called Mountain Glen. Fifteen minutes later, we were parked again, waiting for a new list.

"Chief, uh, Pump 43," the CB crackled. They had just raced north from San Diego. "We are at the address command sent us to. There's really no area we need to spray here. Just wanted to get an update on where you want us to be."

Chief Sam's face hardened. "Okay. You're talking to the wrong guy. Command tells you where to go. I don't have a priority list, Todd. Do you have command's phone number?"

"Copy that, Chief. They advised to contact you, but I'll contact them."

We parked at a spot overlooking the wash and watched men and women from the Los Angeles Fire Department attack the fire. Some had hoses, some shovels. Their faces and jackets were smeared with soot. In the distance, a six-person hand crew marched single file across the charred valley. A woman from the Los Angeles *Daily News* began interviewing Chief Sam as he sat in his truck, so I got out and wandered over to join George, arriving just in time for one of the LAFD firefighters to approach the window.

"So, do you guys just do a certain area?" the firefighter asked. "You go to certain addresses, and if . . . ?"

"If they're in danger, we try to go ahead and spray, yeah," George said. "We try to get ahead of it, but with erratic winds like this, you know . . ."

"Yeah," said the firefighter. He knew.

"It's like the old days," George's partner offered, "with all the insurance companies."

It was an obscure—if accurate—reference to London in the seventeenth century, when any firefighting was done by private insurers. The firefighter took it in. After he left, George rolled up the window. "You see?" he said. "He feels better now."

I returned to Chief Sam's truck, and we drove around the corner, out of the smoke. The smooth jazz came on again, and then he turned it down to make another call: "Hi, I'm a fire chief here in L.A. I was just at the Hyatt, and could you send me some of those wasabi peas? . . . Which one is it? With the green ones? . . . The Hyatt . . . Okay . . . And how much would it . . . ? Yeah. Could you send me a box of those? A big box? . . . Okay. Send me three pounds."

GLOBAL WARMING POSED a grave danger to insurance companies, but it was also something else: free advertising on a biblical scale. Increased risk was a problem only if it wasn't hedged or somehow priced in. Otherwise, it was a business opportunity. Simon Webber of the Schroder Global Climate Change Fund told me that Munich Re—the world's largest reinsurer, with thirty-seven thousand employees in fifty countries and as much as $5 billion in annual profits—was his top holding. A rival manager, Terry Coles of the F&C Global Climate Opportunities Fund, explained how a good hurricane season helped insurance companies hike rates. "People often expect it to be a big negative for insurers," he said. "You get a big sell-off of stock. But unless a really serious one comes through, they'll put the premiums up and actually get the benefit of improved margins."

In 1992, when category 5 Hurricane Andrew struck Florida and Louisiana, insurers paid out more than $23 billion in claims—$1.27 for every dollar of premium collected that year. They turned to catastrophe-modeling companies such as Eqecat and Risk Management Solutions (RMS)—the quants of the insurance industry—which used a century of weather data to predict future losses, and then they raised premiums accordingly. In 2005, after Hurricane Katrina, the first category 5 storm of the new climate era, they paid out more than $40 billion but, thanks to an expanded market and better models, only 71.5 cents per dollar collected. That year, the industry still made $49 billion in profits. It has profited, sometimes more, sometimes less, every year since. After RMS updated its hurricane model in 2006—by flying four scientists to a vacation spot in Bermuda for what it called "expert elicitation"—Allstate used the non-peer-reviewed results to justify jacking up rates in Florida by 43 percent, a move blocked by state regulators. State Farm was similarly blocked from hiking rates by 47 percent. Instead, both dropped tens of thousands of policies, as Allstate also did in storm-surge-threatened areas of New York long before Hurricane Sandy arrived: thirty thousand canceled policies in the five boroughs alone. The companies had more success raising rates in California: a few weeks after the Little Tujunga fire, the state insurance commissioner would approve what amounted to $115 million in increases for State Farm and Farmers—hikes of 6.9 and 4.1 percent, respectively. Allstate's 6.9 percent hike was approved in January 2009.

The once staid insurance industry seemed ripe enough for growth that even Silicon Valley was jumping in. In 2006, a Berkeley grad had founded what would become the Climate Corporation, which harnessed the power of big data—climate modeling, hyperlocal weather forecasts—to sell crop insurance to farmers in the Midwest and eventually weather insurance to the whole world. By 2011, it had compiled fifty terabytes of raw data and raised more than $60 million from backers including Google Ventures; Allen & Company; the Skype founders, Niklas Zennström and Janus Friis; and the green-tech kingmaker Vinod Khosla, who said it would "help farmers globally deal with the increasingly extreme weather brought on by climate change." Its CEO claimed that $3.8 trillion of America's GDP and 70 percent of its businesses were affected annually by the weather; he'd come up with the idea for the Climate Corporation while commuting to his former job at Google. His route took him past a beachside bike rental place: open and bustling when sunny, shuttered and financially underwater when rainy. As the Climate Corporation found new ways to underwrite customers' bets, its own bets were underwritten by the traditional reinsurance industry, the biggest source of funding for climate science outside governments. "If we have a loss, the reinsurer covers 100 percent of the loss," the CEO told a crowd at Stanford. "I mean, venture capitalists don't want to be betting on the weather. They want to bet on a team that can help other people's capital bet on the weather."

Growth was everywhere for the innovative. AIG had Firebreak, but where rich policyholders were clustered on risky coastlines, its Private Client Group was offering the Hurricane Protection Unit: men with GPS units and satellite phones who were on the scene after a storm blew through, boarding up broken doors and windows, patching holes in roofs, covering skylights with tarps, evacuating valuable artwork. In the corporate world, Munich Re's Kyoto Multi Risk Policy protected investors from carbon-credit defaults, and its weather derivatives helped solar projects hedge against cloudy days, wind projects against calm

days. Munich Re would host a "climate liability workshop" at Princeton
a day after Little Tujunga. (It might be prudent to phrase policies so as to
limit climate liability, they determined.) At climate talks in Poznan and
Copenhagen, the reinsurer pushed its adaptation plan for the developing world: a $10-billion-a-year insurance pool funded by governments
but run, of course, by Munich Re. Two other reinsurers, the U.K.'s Willis
Group and Bermuda's RenaissanceRe, were meanwhile pouring money
into hurricane research, the latter also into hurricane modification:
weakening storms by seeding the clouds with aerosols or particles of
carbon. In July 2008, after the Inuit of Kivalina sued the energy companies, Liberty Mutual introduced the world's first insurance policy to
protect corporate executives from lawsuits "stemming from the alleged
improper release of carbon dioxide."

A NEW FIRE appeared just before lunch. Chief Sam spotted it from
Gavina Avenue: a plume of dark smoke rising somewhere to our west. A
new crisis, a new opportunity. A real purpose, perhaps, for his men. But
because we weren't tied into the public system, crucial minutes passed
before our dispatchers could tell us the fire's name, Oat Mountain, or its
location, above the Porter Ranch neighborhood, which was a dozen
miles away. Even after they got it—from the Incident Page Network, an
$8.95-a-month public-alert system—they could tell us little about its
size or trajectory. Thankfully, the public firefighters, racing downhill
with their sirens on, seemed to know where to go. "You see that fire right
there? Where all these units are being rerouted?" Chief Sam asked the
crew of Pump 43. "You guys start heading in that direction." He directed
the two AIG pumps to follow. Porter Ranch was "an upscale community," he said. "It's high volume with AIG clients."

Before we joined them, we stopped for tacos. The Ranchito was a
small Salvadoran place in a strip mall, below the Little Tujunga police
line, next to a Rite Aid. Everyone clapped when we walked in. "Are we

doing a good job?" Chief Sam asked. He and George ordered two fish tacos apiece, and we sat down facing the television. When there was a signal, KCAL 9 had live footage of smoke and flames, aerials of burning foothills. A ticker gave updates: Porter Ranch was under mandatory evacuation. Hundreds of firefighters were converging on the area near Sesnon Boulevard. We were captivated. This was the best intel yet.

Mayor Antonio Villaraigosa of Los Angeles was about to give a press conference. "I'll bet he'll be wearing his jacket again," Chief Sam said. He was right. The mayor stood in front of the microphones wearing a yellow fire jacket just like the one Chief Sam was wearing. He began talking: "I'd just like to say that all the Los Angeles firefighters, all the firefighters . . ."

". . . and Firebreak . . . ," said Chief Sam under his breath.

". . . are doing a great job."

We stayed for forty-five minutes, soaking up information. As we left, the taqueria's owner stopped us. "Do you guys want some bottles of water or anything?" he asked. We were all set, George told him.

We had barely stepped into the parking lot when a woman came running up, begging for help. Suddenly we were all running into the Rite Aid. A girl in line for the register had collapsed. She was lying on the floor, surrounded by a crowd. George kneeled down, speaking to her in Spanish. She was having an asthma attack, he said. He was about to move her when three LAFD firefighters, responding to someone's 911 call, rushed through the door. Three men in yellow looked at three other men in yellow. There was an awkward pause. "We're gonna let you guys have it," George said, and we left to find the fire.

IF FIREFIGHTING CAN be thought of as a rough metaphor for fighting climate change, then public firefighting would be closer to mitigation—cutting emissions for the good of all—while private firefighting would be more akin to adaptation, with individual cities or countries endeavoring to protect their own patches. It is worth remembering that in the case of firefighting we have tried this before.

The man who gave Britain its first firefighters and the world its first fire insurance was a Puritan preacher's son who was born with the name If-Jesus-Hadn't-Died-for-Thee-Thou-Wouldst-Be-Damned Barbon. He later went by Nicholas. In 1666, when he was in his late twenties, the Great Fire of London was lit. It started in a baker's oven on Pudding Lane—someone overcooked some bacon—and because the baker's house was made of wood and his neighbors' houses were made of wood and London had no firemen, it spread easily. People ran in every direction, carting away valuables in horse carriages. "The noise and cracking and thunder of impetuous flames," wrote one observer, "the shrieking of women and children, the hurry of people, the fall of towers, houses and churches, was like a hideous storm." Two prisons, eighty-seven churches, and more than thirteen thousand homes, housing seventy thousand of the city's eighty thousand citizens, were destroyed.

Economists now remember Nicholas Barbon as one of the world's first free-market philosophers. His writings were born out of the Great Fire. With so much land now cleared and available for cheap, one of his responses to it was to become a developer—"the leading speculative builder of his generation," according to the historian Leo Hollis. Barbon's 1685 pamphlet, *An Apology for the Builder,* was written to protest Britain's new building tax and protect his business. It celebrated what happens when people cluster in cities: "Man being Naturally Ambitious, the Living together, occasion Emulation, which is seen by Out-Vying one another in Apparel, Equipage, and Furniture of the House; whereas, if a Man lived Solitary alone, his chiefest Expence, would be Food." Barbon railed against government meddling and called building "the chiefest Promoter of Trade." He would have liked Southern California before the crash.

In 1690, nearly a century before Adam Smith described the invisible hand, Barbon followed up with his most famous work, *A Discourse of*

Trade: "The Native Staple of each Country is the Riches of the Country, and is perpetual, and never to be consumed; Beasts of the Earth, Fowls of the Air, and Fishes of the Sea, Naturally Increase: There is Every Year a New Spring and Autumn, which produceth a New Stock of Plants and Fruits. And the Minerals of the Earth are Unexhaustable; and if the Natural Stock be Infinite, the Artificial Stock that is made of the Natural, must be Infinite." Barbon believed there were no fundamental limits to supply, no real consequences to growth; man could skim infinitely off the top of nature without being subject to its rules. "This sheweth a Mistake," he wrote, by those who would commend "Parsimony, Frugality, and Sumptuary Laws, as the means to make a Nation Rich." What ruined an economy was overregulation. What made an economy great was the demand side—spend, spend, spend, grow, grow, grow—and the more and more a man wanted, the more and more he would receive.

More than three hundred years later, in 2005, Maurice "Hank" Greenberg was forced out after thirty-seven years as CEO of AIG, and his aides created a think tank out of whole cloth and installed him as its figurehead. It was part of a larger rehabilitation campaign that would include hiring well-known academics—the dean of MIT's Sloan School, professors from Wharton and the University of Chicago—to say good things about Greenberg, write papers that underscored his free-market genius, and host conferences that let him be keynote speaker. Its name: the Barbon Institute.

Today, libertarianism of the kind Barbon prefigured is strongly associated with climate denial. If fixing the problem could entail more government intervention, some would prefer to deny that the problem exists. But as this position becomes ever less tenable, some groups appear to be making a tactical shift: Rather than deny the science, they deny that mandated carbon cuts are the solution. For the intellectual descendants of Barbon the free marketeer, one philosophically consistent stance on climate change is to champion market solutions such as

Firebreak—which is to say champion the other response Barbon had to the Great Fire.

In addition to building his building company, Barbon had built a firefighting company, the Fire Office. One contemporary described it as "servants in livery with badges, who are watermen and other lusty persons." They were always ready "when any sudden fire happens, which they are very laborious and dexterous at quelling, not sticking in cases of necessity to expose themselves to great hazards." The Fire Office offered insurance policies for seven, eleven, twenty-one, or thirty-one years—two shillings, six pence per pound of rent for a brick house, twice that for a wooden one, with the services of the lusty watermen included. Barbon signed up more than four thousand clients.

But the company soon attracted competition: the Friendly Society, the General Insurance Company, the Hand-in-Hand Company. Each brigade had its own uniform—blue coats with red linings, or blue shirts with silver buttons, or yellow pants and silver-buckled shoes—and its own fire marks, metal plaques posted on homes so that everyone would know exactly who should save whom. Whenever part of London burned, the brigades fought so fiercely against one another for water and staging areas that authorities had to impose fines: five shillings for hitting a rival fireman; two shillings, six pence for pouring water on him. It was chaotic and, in fact, inefficient. By the early nineteenth century, the disarray was such that private firefighters were replaced by public firefighters, for whom the only adversary was fire.

Seen up close, Chief Sam and his men also seemed less efficient than the public crews. Like the exotic financial instruments that had sunk AIG, like so many responses to the effects of carbon emissions, like so many Band-Aids when the cure was to never have gotten hurt in the first place, the Wildfire Protection Unit was a solution so complicated that the root problem had become an abstraction. Public firefighters fought fires. They discovered where they were burning, went there, and tried to put them out. But Chief Sam and his men did something more complex. In real time, using the best communications systems they could afford without cutting into profits, they got clients' addresses from dispatchers in another state, who had to get them from AIG and Farmers reps in yet another state. They had to figure out where the fire was going. They had to figure out which client addresses were in the way. They had to get to them, even if it meant sneaking across police lines. When they reached an address, and if the house was in fact in danger—a rarity, I was learning—they had to jump out of their trucks, unroll their hoses, spray their Phos-Chek, retrieve their hoses, jump back into their trucks, and race to the next threatened address on their list. Unless, that is, the fire had changed course and the list needed to change, too. When that happened, they had to start all over again. The libertarian dream was a logistical nightmare.

THE HOME CHIEF SAM SAVED, the one I can verify myself, was on Andora Avenue in Chatsworth—"Pandora," he called it, until command straightened out the list. It was ranch-style in the fullest sense: sprawling, with horse stables out back and a twenty-foot American flag up front, set amid a series of equestrian estates in Chatsworth. To reach it, we went against the tide of a full evacuation. Children walked horses and mules down the middle of Topanga Canyon Boulevard. A man in a cowboy hat led a pony half his height. SUVs towing horse trailers clogged intersections. We took a detour down Chatsworth Street and passed more horses being loaded into more trailers, a sports car being loaded into another trailer, a man throwing bags into the trunk of his blue Jaguar, a sign for a missing dog. Public firefighters were nowhere to be seen. The sky was turning yellow. The flames were minutes away.

The homeowner was standing at the entrance to her estate, loading her white pickup with the engine running. Her three young daughters sprinted out of the house. All were wearing surgical masks.

"We're with AIG, your insurance company," Chief Sam told the woman.

"Oh, yay!" she said. "We love you!"

Chief Sam directed Pump 23 into the driveway: "Back it in! Back it in! Watch out! You guys, get out of the way!" It inched the rest of the way in, and the men unspooled the hose. A hot wind whipped the flag back and forth. We sprinted toward the stables. They began to spray the shrubs behind the white guesthouse, the estate's first line of defense. Then the walls of the guesthouse itself. The wooden arbor alongside the guesthouse. The corral. The brown tack room. The blue Adirondack chair. The rear deck of the main house. The patio of the main house. The patio furniture's cushions. The roof and walls and windows facing the brick back courtyard, next to the leaf-choked pool. The wooden furniture at poolside. The roof of the gazebo. The trunk of a palm tree. The roof of the four-car garage.

Chief Sam ran with the crew the whole time, slinging the hose over his shoulders to get it around corners, shouting directions, pointing. When they did it wrong, he did it for them, wresting away the nozzle and blasting the walls until they dripped. The rush ended after twenty minutes, and we gathered on the lawn, breathing hard.

A neighbor appeared. She seemed to have mistaken Chief Sam for a public firefighter, and all of us attempted to play along. "You could get right to the fire at the second property down, Tres Palmas," she said. "They've got double gates all the way."

"Okay, okay," Chief Sam said. The air filled with smoke.

"I've got trails in here," she said. "You can pull all the way in." She pointed down the street, toward the flames, waiting expectantly.

"Okay," Chief Sam said, barely looking at her. "We've got more resources coming."

SIX

UPHILL TO MONEY

WHERE WATER RUNS WHEN IT RUNS OUT

The offices of the world's first water-focused hedge fund were down the Pacific coast from Little Tujunga, in the sprawl of greater San Diego, where they overlooked a parking lot near two malls and four Starbucks in a subdivided landscape developers call the Golden Triangle. The neighborhood gets its name from the isosceles pattern of intersecting freeways that bounds it on three sides. It gets its water from the nearby Alvarado Water Treatment Plant, run by San Diego's Public Utilities Department, which gets its water from the larger San Diego County Water Authority, which gets its water from the yet larger, Los Angeles–dominated Metropolitan Water District, which gets much of its water—in greater proportions during California's droughts—from the 242-mile Colorado River Aqueduct, which gets its water from a reservoir straddling the Arizona border, Lake Havasu, which gets its water from the fourteen-hundred-mile Colorado River, which gets its dwindling water from thousands of streams, snowfields, lakes, and springs in a drainage basin covering nearly 250,000 square miles across seven western states. Without imported water, the 2.7-million-person San Diego metropolis, like much of Southern California, would be no more capable of supporting people than the coastal desert it once was. It was well after Chief Sam's fires had been quelled when I visited the region

again, at the height of a hot summer in 2010. I got my water that morning from a secretary, who handed me a tiny plastic bottle—Poland
Spring, was it?—from the fridge.

The man I'd come to meet was John Dickerson, the founder and CEO
of the sixteen-person Summit Global Management, a former CIA analyst, and the buyer of billions of gallons of water in two vital, desiccating
river systems I would spend weeks tracing: the Colorado and Australia's
Murray-Darling. Both systems had experienced unprecedented recent
droughts, which scientists tied to climate change. Financial managers
like Dickerson, meanwhile, had experienced the opposite: a flood of
money. Summit had launched its first water fund in 1999, but "for a long
time," he told me, "I was a voice in the wilderness. We couldn't get anybody to buy our fund. Then came Al Gore and his stuff, the whole
global-warming thing, droughts. Water became the go-to idea."

For the climate investor, water was the obvious thing. Carbon emissions are invisible. Temperatures are an abstraction. But melting ice,
empty reservoirs, lapping waves, and torrential rainstorms are physical,
tangible—the face of climate change. Water is what makes it real. After
An Inconvenient Truth, during 2007's record melt in the Arctic Ocean,
at least fifteen water mutual funds had launched globally, more than
doubling the number in existence. In two years, the amount of money
under management ballooned tenfold to $13 billion. Credit Suisse, UBS,
and Goldman Sachs hired dedicated water analysts, the latter calling
water "the petroleum of the next century" and referring to "major multi-
year droughts" in Israel, Australia, and the American West. "At the risk
of being alarmist," Goldman said in a 2008 report, "we see parallels with
Malthusian economics."

Citigroup's chief economist, Willem Buiter, would take it further. "I
expect to see in the near future a massive expansion of investment in the
water sector," he wrote, "including the production of fresh, clean water
from other sources (desalination, purification), storage, shipping and
transportation of water. I expect to see pipeline networks that will

exceed the capacity of those for oil and gas today. I see fleets of water tankers (single-hulled!) and storage facilities that will dwarf those we currently have for oil, natural gas and LNG. There will be different grades and types of fresh water, just the way we have light sweet and heavy sour crude oil today. Water as an asset class will, in my view, become eventually the single most important physical-commodity based asset class, dwarfing oil, copper, agricultural commodities and precious metals." It was Etan Bar's elevator pitch from Israel all over again—if there's a market for anything, it's water, because there is no water!—only this time there was real money behind it.

In his office, Dickerson, in his late sixties, sat in a leather chair next to a window and an old PC, and I sat across his desk from him. The walls had photographs of Alaskan glaciers and Utahan deserts—Dickerson took them himself—and a bookshelf had three copies of *An Inconvenient Truth* and two of *Cadillac Desert,* the seminal 1986 book on water and political power in the American West. Its author, the environmental icon Marc Reisner, was a Summit board member before his death in 2000. "There are all these Zen-like things about water," Dickerson told me. "It's the most necessary of all commodities. You know, there's no substitute for it at any price. And we cannot make water. Did you ever think about that, really? Hydrogen and oxygen. You can't grow it. It's a substance that is gonna be forever fixed on this planet."

Where the water was, already the people often were not. "We still have the exact same amount in our ecosphere," he continued, "but the ultimate effect of global warming is that percentage that is freshwater is getting smaller, the percentage that is salt water is getting larger, and the maldistribution of freshwater is getting much more severe." There were record floods in China, unprecedented droughts in Australia. "We seem to go to these climatic extremes," he said. The "supply/demand imbalance" for water—fueled by population growth, accelerated by carbon emissions—was only increasing. It was a situation ripe for speculation, except there was no easy way for most investors to get in. "If you're a mom and pop, sitting in Peoria, so to speak, you can buy wheat futures, pork bellies, oats, orange juice futures," Dickerson explained, but "oddly enough, not water futures."

Summit's first water fund, which was up 200 percent after its first decade, with $600 million under management, had navigated this road-block by picking stocks within the convoluted $400 billion field Dicker-son termed "hydrocommerce": the business of storing, treating, and delivering water for use in households, manufacturing, and agriculture. Dickerson's newer rivals, including funds from Pictet, Terrapin, and Credit Suisse, mostly did the same. They bought shares of builder-operator multinationals such as France's Veolia, of the tagline "L'envi-ronnement est un défi industriel"—"The environment is an industrial challenge"—and Suez, its compatriot and major rival in water treatment and desalination. They bought diggers of ditches, layers of pipelines, and manufacturers of filters, pumps, meters, membranes, valves, and electronic controls. They also bought privatized utilities in cities big and small, though these, despite—or perhaps because of—widespread fear of financialization of water, serviced just 12 percent of the American public, 10 percent of the globe, and they could hike rates only as high as regulators would allow.

The investable universe of hydrocommerce was small—about four hundred publicly traded companies, according to Summit—and the recent gush of new investors was dramatic, and prices had become inflated. "I'm a value investor," Dickerson said. Rather than try to time the ups and downs of the market, he bought stocks strictly based on their price in relation to what the company itself was worth. So in over-heated 2007 and 2008, he found himself selling more than buying—and the resulting cash hoard was one reason Summit had survived the financial crisis largely unscathed.

Dickerson's second response to creeping global drought and sudden pressure from rival investors was more intriguing and, as other funds again began to follow his lead, more significant: He had decided that hydrocommerce—things merely related to water—wasn't enough. He wanted actual water: "wet water," as he called it. Raw water. The thing itself. In June 2008, he opened a second hedge fund, the Summit Water Development Group, to amass water rights in Australia and the American West. Already the new fund had attracted hundreds of millions of dollars. "I've watched water rights go up and up," Dickerson told me. "Just tick, tick, tick, tick." He lifted a hand in the air, jerkily raising it higher and higher. "The real future," he said, "is going to be the direct assets—not through the medium of a utility, not through the medium of a pump company—but the direct, physical water assets."

"WATER FLOWS UPHILL to money," wrote Marc Reisner in *Cadillac Desert*. The adage captures the spirit of the moment, but as a description of what is happening to water the more the world warms, the more rivers like the Colorado wither, it is imprecise. Water is heavy—about 8.3 pounds per gallon—and to move it in bulk without significant help from gravity or the Army Corps of Engineers was still too expensive for privateers to profitably pull off. If Citigroup's Willem Buiter was right about the future of international water markets, parts of that future were proving slow to arrive.

The comparative shrewdness of zero-sum strategies like that of the Summit Water Development Group, which bought up rights within a stressed river system rather than importing water from overseas, is illustrated by the failures of recent bulk-shipping schemes. In the Mediterranean, a $150 million water treatment and export facility was completed in 1998 near the mouth of Turkey's southerly Manavgat River, and four hundred miles south an intake pipeline was built in Ashkelon, Israel, now home to the massive desalination plant I visited with IDE. The two countries' 2004 "water for arms" deal—certain high-tech weapons would go to Turkey, 13 billion gallons of Manavgat water a year to Israel—quickly fell apart over high costs and diverging politics.

In Sitka, Alaska, one company after another signed one-cent-per-gallon contracts for up to 2.9 billion gallons a year from the city's Blue Lake. A pipeline was built in 2007 to fill tanker ships with 60 million gallons apiece, and the Texas-based S2C Global Systems, the latest lessee, declared that it was creating its first "World Water Hub" in southern India. But the contracts in Sitka kept requiring extensions; especially in India, the water-poor are often money-poor, and S2C has seemed unable to find the right buyer. Not a drop has been shipped from the harbor.

In Iceland, per capita the most water-rich country in the world, at least until Greenland gets its independence, three successive ventures have negotiated leases in which they would pay a tenth as much per gallon as S2C for the water running off the Snaefellsjökull, the volcano from Jules Verne's *Journey to the Center of the Earth*. One was a fraudulently run hedge fund managed by a Canadian former dentist named Otto Spork, another a more aboveboard venture by a British hedge fund called Moonraker. None has exported more than a few bottles, let alone turned a profit. In 2011, I visited Spork's half-built water plant on the flanks of the volcano: sheet metal, dirt floors, 100,000 square feet, two pipelines dumping ninety liters per second of glacier water uselessly into the sea. "Everywhere in the world," explained one Icelander who had run the numbers on shipping costs, "it is cheaper to do desalination."

The water industry's biggest dreamers are the "bag and drag" men: those who would fill enormous polyester bags with freshwater and tow them through the oceans. Best known is Terry Spragg, the obsessed inventor of the Spragg Bag (www.waterbag.com). In the early 1970s, the Rand Corporation was studying how to tow icebergs to perennially water-stressed Southern California, and Spragg, who was ski bumming through the Rockies at the time, got in touch with the authors. Soon he was representing Prince Mohamed Al-Faisal of Saudi Arabia, the founder of Iceberg Transport International Ltd. and sponsor, in 1977, of the First International Conference and Workshops on Iceberg Utilization for Fresh Water Production, Weather Modification, and Other

Applications. (Papers and talks included the succinctly titled "Calving.") The next year, Spragg got the California legislature to endorse iceberg towing. But gradually he lost faith in it himself. Icebergs melted too quickly. "I said: 'Let's just go to the mouth of a river and fill a bag,'" he told me. "I'm just trying to solve a problem: There's enough water in the world, just not in the right places."

Spragg made his first water bag in 1990 and tested it in Puget Sound near Seattle. He filled it up at the edge of the rain forest on the Olympic Peninsula—"the best place in the whole United States for water"—and began to tow, then watched the bag split open and spill 700,000 gallons into the sound. Undeterred, he had an MIT professor help him design and patent a zipper system with big enough teeth to hold two big water bags together. He hired the Colorado engineering firm CH2M Hill—ubiquitous in the profitable battle against drought, with tunnel projects near Las Vegas and desalination plants in Australia—to design a water bag loading and unloading system. He began envisioning bladders the shape and general size of a nuclear submarine, zipped one to the next in fifty-bag trains and deposited one by one in depots worldwide. In 1996, he completed a successful drag across Puget Sound to Seattle, only to have a tugboat run into his docked prototype. He had no insurance.

Spragg took interest in the Manavgat, hiring an agent in Israel, as he would later do in Australia as the Murray-Darling ran dry. He wrote a heroic novel about saving the Middle East via water bags—*Water, War, and Peace*—that starred a thinly autobiographical character named Gerald Earl Davis. But the Manavgat still flows mostly to the sea and the novel remains unpublished, and for two decades Spragg has been trying to raise money for another prototype. As for the common petroleum-to-plowshares dream of shipping water with old single-hull oil tankers—which are in low demand after the *Exxon Valdez*—he and other experts are dismissive. Single-hull tankers may be inexpensive to buy, but they are expensive to retrofit. Not only do the ships' holds need to be cleaned, but pipes, pumps, valves, and washers need to be replaced.

"I've seen the numbers," says Spragg. "Basically, it's cheaper to take the tanker and cut it down and use it for scrap than to redo it."

Another would-be bagger was Ric Davidge, a deputy to Secretary of the Interior James Watt during the Reagan administration who later became Alaska's first director of water. In 2000, he enraged Northern Californians with a proposal to bag water from two Mendocino-area rivers, the Albion and the Gualala, and tow it six hundred miles south to import-dependent San Diego. The plan had to be abandoned in the face of angry opposition, and Davidge had to rename his company. At the time he was also chairman of a consortium called World Water SA, which consisted of a large Japanese shipping line, a Saudi industrial conglomerate, and a Scandinavian water bag company, Nordic Water Supply. Nordic's bags were among the few in history to see commercial use, delivering five million gallons a pop from the Manavgat to arid Cyprus. Bags of freshwater float high in the salty Mediterranean, but Nordic's costs were more crippling than Davidge and the other partners were told. Nordic went bankrupt soon after Davidge was run out of Mendocino. The "first law of Davidge water," he explained to me, "is that everybody lies about transport costs. Don't talk to me about sources. I know sources all over the world. Talk to me about conveyance systems."

A decade later, Davidge had mostly given up on water bags. There were promising new tanker designs coming out of Europe and Asia, he said, and his new company, Aqueous, had lately been negotiating with Sitka. Spragg, meanwhile, had not given up. "In Spragg's perfect world," he told me, "which may be crazy, I could store the bags on a big spit on the Olympic Peninsula, then take them out into the ocean and let them go, track them by GPS. The currents will take them all the way to Southern California."

A HUNDRED MILES east of San Diego, I witnessed the lengths to which the city was already going to secure more water supplies, how it was getting by until the tankers and water bags appeared. The first sign that something was odd was near the dusty town of Imperial, California, where there was lettuce growing in the desert. As I drove along Interstate 8, the lettuce gave way to cabbage. The cabbage gave way to alfalfa. "Now you see where all that water goes," my host, an engineer named Todd Shields, said. We continued east, and soon enough it all gave way to sand once again. Just south of the freeway and just north of the new border fence with Mexico was the All-American Canal: the biggest irrigation canal on the planet, California's first and biggest claim on the waters of the Colorado River, and, most recently, the site of the biggest bulk-water deal in history—a measure of the market that men like John Dickerson were chasing.

In 1899, the Canadian American entrepreneur George Chaffey, an irrigation genius who had founded the "model colony" of Ontario, California, before being recruited to Australia to duplicate his efforts on the banks of the Murray River, had begun staking out his latest venture, Imperial. Chaffey's publicist sold settlers on the image of the Imperial Valley as the Egyptian delta, the Colorado River as the Nile, themselves as Joseph and the chosen people—pilgrims, not just pioneers. On May 14, 1901, near a volcanic outcropping called Pilot Knob, the wooden Chaffey Gate began diverting the waters of the Colorado through a series of trenches and canals toward Imperial. Water rights being a matter of seniority, not geography, it may have been the most important moment in California's history—the start of its claim to a river whose flows come entirely from other states and the start of a system of aqueducts that have allowed its cities to flourish where no city should rightly be. Under the 1922 Colorado River Compact between seven western states and Mexico, California is apportioned 4.4 million acre-feet of water a year, or about 1. 4 trillion gallons—the most of any state in the system, more than a third of what's promised to the rest. If there's any surplus above the 1.5 million acre-feet committed by treaty to flow onward to Mexico, California takes much of that, too. Thanks to the All-American, which got its name after an earlier canal was rerouted in the 1930s to stay wholly north of the border, the obscure Imperial Irrigation District (IID) now controls 20 percent of the Colorado's flows. Imperial receives 2.92 inches of rain a year, a third of what San Diego gets. The surrounding desert, once known as the Valley of the Dead, has become one of America's prime farming regions. Two-thirds of the country's winter fruits and vegetables are grown here.

In 2003, as the Colorado fell into drought, the IID agreed under threat of federal intervention to sell a record 277,000 acre-feet of All-American water—equivalent to 90 billion gallons, 5,000 Panamax tankers, or 20,000 Spragg Bags—to the San Diego County Water Authority. Most of the new All-American water would go to the City of San Diego, where until 2012 voters kept rejecting the idea of bolstering their meager water supply with treated sewage. Most of the city's water, if history was any guide, would go to keeping four hundred parks and golf courses green. Residents themselves would dump half their water into their yards. In poorer, politically weaker Imperial, farmers would idle tens of thousands of acres of cropland and take generous water-rights buyouts, some happily, many grudgingly. For me, the All-American Canal symbolized the essential truth of Marc Reisner's adage about water and money, along with its corollary in a warming world: Shit rolls downhill.

Todd Shields, my host, was managing the most controversial part of the record water deal. For most of a century, the All-American Canal had been lined with dirt, and at least twenty-two billion gallons—enough for 122,000 American households—were lost each year through its porous walls. The IID would now line the canal with concrete, and San Diego would pay the $290 million in construction costs. The problem was that for most of a century the lost water had slipped under the Algodones Dunes, ignoring the international border, and percolated up in the Mexicali Valley. Farmers there had used the leakage to turn Mexicali into one of their country's largest producers of alfalfa, asparagus, scallions, and cotton. Take it away, and soon hundreds of people would be out of work, tens of thousands of people would be out of potable water, and sensitive wetlands would be drained. Lawsuits against the canal upgrade had been filed by environmental groups, the city of Calexico, California, and the Mexicali Economic Development Council. But they'd been trumped by a provision slipped into the final lines of a 279-page tax bill in the final hours of the final congressional session of 2006. The edict—ignore environmental-assessment requirements and, "without delay, carry out the All-American Canal Lining Project"—was the doing of three Colorado River system senators: Dianne Feinstein of California, Harry Reid of Nevada, and Jon Kyl of Arizona. "In a time of increasing population and decreasing water supplies as a result of global warming," said Feinstein, "I believe it is critical to save every drop of water."

I watched the border fence rise and fall as Shields updated me on his four-hundred-person crew's progress: twenty-three million cubic yards of sand and soil moved, eighteen of twenty-three miles nearly done, thirteen months to go. We passed an ad hoc roadblock stopping traffic in the opposite lane: orange cones and white SUVs with green bands marked "Border Patrol," a cluster of officers looking for illegals. Since the All-American's 1942 completion, there had been almost six hundred documented drownings of undocumented migrants—almost one a month—and workers regularly collected the dead from the pools above the hydropower plants. "There's a trash rack before the water goes through," Shields explained. "You don't want debris to go into the turbines. That's generally where they find the bodies."

Near where the freeway crossed the canal, not far from Pilot Knob, Shields and I pulled over. Workers had dug an entirely new trench in this section, and there was a waterless, rectilinear canyon 150 feet across and nearly 100 feet deep, swarming with men and machines. Otherworldly "jumbos"—inclined platforms that stretched top to bottom on the far bank—were creeping east on metal tracks. One had four giant blue spools of plastic joint material and a vertical sluice that brought down endless streams of wet concrete. Another had a team of eight men in hard hats and blue jeans who buffed the concrete, leaning forward on what looked like industrial mops, straining in the sun. In this heat, the concrete would be dry within thirty minutes, and then another jumbo would grind past and spray sealant, and the bank would go from gray-brown to gleaming white. In about a month, Shields told me, they would cut the weirs at both ends, and the new section of canal would flood with water. He had donned a yellow safety vest over his flannel shirt, and now he stood at the edge of the void and watched his men work. They were speaking to one another in Spanish.

"This is kind of a special project for me," Shields said. His grandfather Clyde had led the All-American's 1930s survey crews, he explained, and had later worked on the California State Water Project. Shields had been inspired to become a civil engineer himself, but unlike his grandfather he was not technically a government employee: He had been remanded to the IID from Parsons, the Los Angeles engineering firm that had been the first to dream up NAWAPA, the North American Water and Power Alliance, which I had heard about when sitting in on Michael Byers's class at the University of British Columbia. Early in his career, Shields had seen a scale model of it once, had gotten to really study the megaproject that had so scared Canadian water nationalists. "It was just this huge board with all these systems," he said, "interesting as hell to me." He thought it could work. "I'm sure it would work," he told me. "It's technically feasible. You know, it would solve water needs. It would probably violate lots of environmental needs. It's a societal value judgment."

Shields told me he believed the climate could be changing, sure, but he didn't believe humans were causing it, and he didn't believe it was a crisis. "What that whole global-warming fear leaves out," he said, "is that there will also be positive effects."

The next day, I crossed the border. The Mexicali Valley's fields were flat, straight, perfectly manicured, lorded over by the Algodones Dunes yet radiantly green. An American company named El Toro shuttled workers back and forth in school buses, and massive sprinkler systems watered the land. In the city of Mexicali, three blocks from a border wall where the streets abruptly ended, near a florist and a health clinic, I entered a small blue building housing a plaintiff in the lawsuit against the lining project. René Acuña, the director of the Mexicali Economic Development Council, wore a maroon shirt and sat in a leather chair and explained, calmly at first, that Mexicali wasn't some maquiladora town. It had a million people, and its entire population could live for a year on the water taken by the new Canal Todo Americano. A third of its economy was agriculture. "Our prosperity has always been based on water," he said. He showed me a photograph of the clear, filtered water that bubbled up from the border—not too saline, not yet. "But those fields are going to die," he said. "And then see where the people go."

JOHN DICKERSON HAD wanted to make sure I understood how a water-rights hedge fund like his could exist, so he took the time to explain how something like water could be bought. The concept was straightforward: In some parts of the world, a water title was like a land title. The details, on the other hand, were extremely complicated.

"In the United States," he began, "we have two systems with water rights." In the eastern United States, as in most parts of the former British Empire, the courts followed riparian law, part of traditional English common law. "If you have ten hectares of land, you get x liters from the Thames," he said. "If you have a hundred hectares, you get ten times x from the Thames." In states like Indiana, Ohio, Michigan, and Maine, water could not be stripped from the land and sold as a separate commodity.

The West was different. The Homestead Acts let pioneers win deeds to federal land if they lived on it for a period and made improvements. "Wagon trains would leave Missouri, headed toward Oregon," Dickerson said. "They would see a valley, stop, and say, 'This looks good.' They all settled along water." In dry years, common law could not stop farmers living upstream from using up all the water before it reached the others, for none of the settlers yet owned the land they worked. "People living at the lower end of the valleys would go blow up dams," he said. "The government had to start sending federal marshals out." Water courts formed to sort out the mess, and as the deeds came, they came in pairs: one for land, one for water. The basic water law of the West became first come, first served. "First in time," said Dickerson, "first in right." This was why the All-American Canal was so important for California. So long as they had been exercised and thus kept valid—the water was put to "beneficial use," not hoarded—the oldest rights were the most valuable, and they could be freely traded.

"Any water title sold to San Diego or Denver today was first bought off a farmer or rancher way back then," Dickerson told me. He abruptly stood up, walked to the back of the room, and returned with a diagram that he slapped down on his desk. It was a hydrologist's straight-line graph of Colorado's South Platte River, a depiction of tributaries and water rights. "This is just to give you some sense of the complexity involved." It showed a disorderly network of dozens of colored lines—reds, blues, and greens—that converged at odd angles across the page. "Those things are reservoirs," he said, pointing. "Those are ditches. Just look at this stuff. You almost have to have a magnifying glass. See, here's a water right: year 1910, thirty-two thousand acre-feet, so-and-so cubic feet a second. Everything has a date on it. It's incredibly complex." But once you understood the rules, he said, you were a big step ahead of the guys buying utility stocks. "A lot of people say to me, 'John, water is a regulated business. What are you saying here, that water's a free-market good?' And I say, 'Well, no, water is not a regulated business. Water *utilities* are a regulated business.'"

Dickerson planned to eventually take the Summit Water Development Group, his "wet water" fund, public—meaning a way for mom and pop in Peoria to finally speculate on water, plus a big payout for early fund investors who had put up the minimum $5 million buy in. In the meantime, he was playing what he called an "aggregation game." Across the American West, up and down the Colorado River system, Summit took stakes in private reservoirs, in ditches dug 150 years ago by pioneering ranchers—spending $500,000 here, $1 million there—with the goal of accumulating enough water to sell as a package to suburban boomtowns within the basin. Once resold, the ditches' water would be left in the river to be taken up by city pipes.

After the new millennium and the onset of the worst Colorado River drought in memory, the rural-to-urban flow of water—already the bulk of water trades since tracking began in 1987—had doubled in volume. In the Rockies, snow fell as rain. Sometimes, rain did not fall at all. It was a preview of the future, scientists said: Climate models projected a northward shift of the Hadley cell, a planetwide atmospheric system that circulates warm air from the tropics and cooler air from the subtropics, driving trade winds, jet streams, and, crucially, desertification. For the Southwest, eighteen of the nineteen major models predicted permanent drought by 2050, with an average surface-moisture decline of 15 percent—the size of the decline that precipitated the dust bowl in the 1930s. Now as then, there had been a rural exodus. The population followed the water to the cities or perhaps vice versa. There were now more people to eat food, and there were fewer people to grow it.

Developers had been on a water-buying spree until the housing market peaked in 2007, Dickerson told me. "Water went from $3,000 an acre-foot to $30,000 in some locales," he said. Next came the crash—his moment. ("Three or four times," he said, "we have bought water from bankruptcy courts.") Now, with a fracking boom and an equally water-intensive hunt for other unconventional oils, petroleum interests were on a buying spree. In late 2008 in the upper Colorado basin, one of the biggest players, Royal Dutch Shell, filed for the first major water right on the Yampa River, requesting 375 cubic feet per second, or 8 percent of the river's peak springtime flow. (It later withdrew the request after locals protested.) According to one study, energy companies controlled more than a quarter of the upper basin's flow, more than half of its water storage. Farther south, in Texas, the onset of fracking corresponded to the driest year in state history, and ranchers and cities alike were priced out of the water market. To frack a single well can take as many as six million gallons. In 2011, petroleum companies drilled twice as many new water wells—2,232 across the Lone Star State—as they did oil and gas wells. For the Summit Water Development Group, all this was very good news.

In Australia, the catastrophic drought in the Murray-Darling basin, the continent's overstretched twin to the Colorado, was also good news for Summit. But it was not the only reason Summit was buying here, too. The other reason, Dickerson explained, was that Australia had copied the American West's system of tradable water rights in the early 1980s. Then Australia further liberalized its system, creating what has become the planet's freest and most bustling water market. The Colorado's twin in drought was also its twin in free enterprise. In the Murray-Darling, Summit had secured what outsiders estimated to be at least 10,000 megaliters, or 2.6 billion gallons. It cultivated a diversified portfolio, Dickerson said, as with stocks, buying water in various Australian states that flowed to various crops: wine grapes, citrus, cotton, or almonds. He planned to become something other than a short-term speculator: a long-term rentier. Once Summit purchased Australian farmers' water, he said, the firm banked it and leased it right back to them and their neighbors. Returns were already a safe 5 to 6 percent a year. "There's no risk," Dickerson said. "If some guy doesn't pay, we still own the water. It's like you turn off a tap."

In an era of increasing scarcity, many economists argued, the best way to cut our profligate waste of water was to have active water markets. Championed in Australia by the University of Adelaide professor Mike Young, in America by the Hoover Institution fellow Terry Anderson, founder of Montana's "free market environmentalist" Property and Environment Research Center, the idea was that trading means incentives to conserve and use water efficiently, that markets allowed a scarce resource to flow to the highest-value activities. "One of the things governments can start doing," Dickerson told me, "is to allow water to be priced at what it's worth, then create a mechanism by which the rice farmer can sell his water to the wine producer." That Australia's water trading helped one of the planet's great exporters of rice and wheat through its drought was undeniable. On a macro level, the economy survived remarkably unhurt. Also undeniable were the distortions: By 2008, near the end of the decade of drought and historic evisceration of the $35 billion farming sector, Australia's rice production had dropped to 1 percent of normal, its wheat production to 59 percent. That year, what aid agencies dubbed the "global food crisis" led to protests in Egypt, Senegal, Bangladesh, and dozens of other countries. Adelaide's wine industry, on the other hand, was still thriving.

As I headed out the door into the San Diego sun, Dickerson graciously loaded my arms with books and reports, apologizing that he couldn't give me his only copy of *Unquenchable*. At the top of the stack was *Water for Sale: How Business and the Market Can Resolve the World's Water Crisis*, a book published in 2005 by the libertarian Cato Institute, another Koch-brothers-funded think tank. "Some people don't like it," Dickerson said, "but this is what's coming."

TO TRULY UNDERSTAND what was coming, I had to make another trip, to the continent where the future already seemed to be here. In Australia, Summit Global ran its wet-water operations out of Adelaide, a swiftly growing city of 1.2 million people near the end of the Murray River that had near-brackish tap water and a reputation for bizarre murders. At the height of the decade-long drought that locals call the Big Dry—the worst drought to yet hit the industrialized world—I drove there from

Sydney, dropping south to the Snowy Mountains before traversing the continent westward along the dwindling Murray. It rained when I was in the Snowies, and then it didn't rain again, and the land became ever more orange and empty. On the banks of the Murray, river red gums cast their shadows on flats of cracked mud, and on the sides of the highways every other farmhouse seemed to have a "For Sale" sign. The river was so low that its iconic riverboats could not get through the locks. "The whole rural thing is just going to go belly-up," a captain told me in Echuca. "Everyone's going to move to the city, and there's going to be nothing out here. It's going to be like some sort of *Mad Max* movie."

The sellers in the burgeoning water markets, I learned, were family farmers. Small-time ranchers sold to corporate farms or citrus growers or the government; water flowed uphill to cities and vineyards. The biggest purchaser was the federal government, on a $3.1 billion run of buybacks for what it called "environmental flows." Until the drought broke in late 2010—dramatic floods inundated 250 homes—the river did not readily reach the sea, and government warned that this would be the new normal unless there were fewer demands on the overstretched Murray-Darling basin: By 2030, climate change was expected to decrease local rainfall by 3 percent, decrease surface water flows by 9 percent, and increase evaporation by up to 15 percent. Lining up behind the government were Summit and a growing cast of other funds: Australia's own Causeway Water Fund and Blue Sky Water Partners, Singapore's Olam International, Britain's Ecofin fund, and a company called Tandou Limited, which was owned by a New Zealand corporate raider, the American hedge fund Water Asset Management, and Ecofin.

In Adelaide, a young PR manager led me past a row of brokers sitting in velvet chairs and staring at flat-screen Dell monitors, scanning satellite images on Google Maps: the headquarters of Waterfind, the country's largest water brokerage, which had developed its own software platform for trades up and down the Murray-Darling and touted its plan to become a true stock market—the Nasdaq of water. There were

exchange rates for different parts of the river system, he explained—
owing to evaporation and local regulations, a liter in the Murrumbidgee
Valley might not be worth quite the same as a liter in Murray Bridge—
and there were volume caps to navigate; some states were still protectionist about their water, though it was getting better. These were paper
trades, conducted by phone and Internet. Buyer and seller could be hundreds of miles apart. One would turn off his pumps, and the other
would turn his on. The water market in 2008, near the peak of the
drought, was worth $1.3 billion, and it had been growing by 20 percent
a year. Bulk water in Australia is measured by the megaliter, equivalent
to 264,000 gallons. The price of a megaliter fluctuated wildly. "In the
temporary markets last season," he said, "the low was right around $200.
The high was right around $1,200." In general, though, during the
drought, the price went up.

Another day, I drove into the hinterlands of Adelaide with a former
undercover narcotics detective now assigned to the new crime of water
theft. We cruised near the Murray, looking for action, and he told me
about the tools he had: night-vision goggles for stakeouts, aerial surveillance to detect overly green fields. We stopped at a marina to see all the
houseboats stuck in the mud, and he told me about the tools the criminals had: makeshift dams, surreptitious pumps, hoses snaking over to
neighbors' spigots, and frozen carp. The carp were for the wooden
waterwheels meant to measure each farmer's water allotment. Jam a frozen fish into one of them, and it would stop spinning. By coincidence,
this was known as "spragging the wheel." If an inspector came, the carp,
now thawed, indistinguishable from a wild fish, got the blame. That
water theft was being taken so seriously helped make sense of what else
I was seeing in the world's most liberated water market. The idea that
water could be stolen, like the idea that it could be bought and sold, was
predicated on the increasingly accepted idea that it was something that
could be owned in the first place.

"I'm not so much interested in the cause of the climate change,"

Senator Bill Heffernan told me when we met at the Parliament House in Canberra, Australia's hill-ringed capital. "I'm interested in what we're going to do about it." It was the sentiment of the new age, the free marketeer's emerging mantra in two hemispheres—only Heffernan, the right-hand man to the former prime minister John Howard, a wheat farmer, and a kind of futurist for Australia's conservative party, the Liberals, was beginning to doubt that strong property rights and liberalized markets could really save the day.

When the Liberals were last in power, Heffernan had chaired the Northern Australia Land and Water Taskforce, which investigated whether the country's status as an agricultural power could be saved if production and population moved from the Murray-Darling to its underpopulated, water- and land-rich north. He had great hopes for the Cape York Peninsula, a Kansas-size tropical wilderness populated by a few thousand indigenous islanders and aboriginals, some of whom were pushing to turn these ancestral lands into a UNESCO World Heritage site.

"Climate scientists are saying that over the next forty to fifty years, 50 percent of the world's population will become water-poor," he told me. "They are saying that in the region of Asia, our immediate neighbors, there will be a 30 percent reduction of productive land over the next forty to fifty years. The food task will double in that time, and 1.6 billion people could possibly be displaced. Now, if that science is only 10 percent right, we've got a serious problem. One of the problems with this changing planet is, how are we going to manage world order? I mean, the chief commissioner of the Australian Federal Police said last year that the greatest threat to Australia's sovereignty is actually climate change." The north was dangerously close to overcrowded Asia.

Heffernan knew that foreign hedge funds were circling, and he saw their arrival in Australia's water markets in the same protectionist light. "I don't think we can afford for water to become just a speculative commodity," he said. But I was surprised to learn that water speculation wasn't his main concern. As the drought abated, Arabs, Chinese, and other foreign investors were scouring Australia and the rest of the planet for something else: farmland. Like naturalists the world over, Heffernan was alarmed. As he told one reporter, "We are actually redefining sovereignty."

SEVEN

FARMLAND GRAB

WALL STREET GOES TO SOUTH SUDAN

The day we flew to Juba in an old DC-9, the sun was out and the clouds were distant puffs, and all we could see was green: the dirty green of the Nile, the dark green of the mango trees, the radiant green of the uncultivated savanna. The land was flat and muddy and empty, and it stretched forever. "Just look at that shit," Phil Heilberg said. "You could grow anything there."

We went to see the general immediately after landing, Heilberg taking the passenger seat in an aging Land Cruiser pickup driven by his business partner—the general's eldest son, Gabriel—and me sitting in between them. We rumbled down one of South Sudan's only paved roads, passing Equatorian boys on motorbikes, Kenyans tending makeshift kiosks, the cluster of permanent structures constituting downtown, and the fortified offices of the UN Development Programme, then turned into a neighboring compound. It was surrounded by machine-gun emplacements and thatched-roof huts known as *tukuls*, which were homes for guards and wives. The monkey was gone, Heilberg noticed. The guards used to have a monkey. "Where's your monkey?" he yelled as we drove in.

General Paulino Matip, the deputy commander of the Sudan People's Liberation Army (SPLA), was waiting for us in a dirt courtyard shaded by mango trees. He wore a tracksuit and was slouching in a plastic chair in front of a plastic table with a doily on it, flanked by a dozen elders from his Nuer tribe. His face was expressionless. "Ah, Philippe," he said, and he slowly stood to hug him. Gabriel translated the rest: "The only white man who is good."

In the flurry of news accounts and think-tank reports about what activists had begun calling the global farmland grab, Heilberg and Matip were recurring characters: the Wall Street guy and the warlord, the former AIG trader and the most feared man in South Sudan, twin symbols of what would happen the more populations boomed, temperatures rose, rivers ran dry, and food prices—and thus the value of farmland— shot through the roof. In the previous decade, especially after the 2008 "food crisis" that preceded the financial crisis, rich countries and corporations had acquired an estimated 200 million acres in poorer countries—the equivalent of the combined cropland of Britain, France, Germany, and Italy, or almost 40 percent of arable Africa, or every inch of Texas. It was a territorial shift unseen since the colonial days, and it was happening quietly and bloodlessly, behind closed doors. I'd come here because Sudan, along with Ethiopia, Ukraine, Brazil, and Madagascar, was one of the major target countries—and because Heilberg, convinced he was doing the right thing, was unafraid of the attention.

Heilberg's own patch, leased in a late-2008 deal approved by Matip, was nearly the size of Delaware: a million acres. Irrigated by an offshoot of the Nile, it was level and fertile, safe from drought, and largely free from land mines. The deal, if it was valid, had turned him into one of the largest private landholders in Africa, and crammed in his briefcase was a map showing where he now hoped to double his holdings: six blocks to the east and north of his first million acres, close to the border with Ethiopia, outlined in orange marker.

The goal of Heilberg's Juba visit was signatures. He wanted Matip to lean on South Sudan's agriculture minister and on its president, Salva Kiir, of the politically dominant Dinka tribe, to sign off on his farmland deal. People were beginning to whisper about it—that it was illegal, that

it violated the new country's new land law. Heilberg told me the signatures were mostly for show: The farmland was in the general's home
state of Unity, and Heilberg had it because the general and other Nuer
leaders said he had it. But official approval would reassure potential
investors. And President Kiir had promised his signature, Heilberg said.
General Matip had stopped battling his fellow southerners only after the
2005 peace accord that had ended Sudan's twenty-two-year civil war,
the longest in Africa, and set out a path toward independence for the
south, which would come after a 2011 referendum. He had leverage over
the president: the twenty thousand to thirty thousand members of his
Nuer militia, whose integration into the SPLA was still mostly on paper.

Heilberg turned to the general's son. Gabriel, wearing an Armani
jacket and armed with three Nokia cell phones, looked to be in his twenties, but he was actually thirty-four, according to his Myspace page. Or
maybe forty-two, as he later told me.

"Gabriel, do you have any cows here?" Heilberg asked. Cows were the
way to a Nuer man's heart.

"No, no cows here," said Gabriel.

"You already moved them to Mayom? How many cattle do you have
in Mayom?" Heilberg asked.

"Many," said Gabriel.

The Nuer elders surrounding the general stood up, taking their chairs
with them. A soldier brought us bottled water and cans of Coke. The
general stayed slouched in his seat, his long arms draped over the back
of the chair, and stared blankly into space. He was sixty-eight, ancient
for South Sudan, a survivor of lifelong war, now suffering from diabetes
and high blood pressure. Heilberg considered him one of the savviest
men he'd ever met. "He's a capitalist," he told me. "All the other guys are
Commies. He understands that if I put money in, I deserve to make a
profit." Heilberg claimed that in famously corrupt Juba, Matip was the
rare leader who didn't demand bribes.

"The south has got to give the general control of the purse," Heilberg said, leaning in. "People have been approaching me. A private security firm—mercenaries—they want to come and train the troops. Even the Israelis, they want to sell weapons and training. They want to know if I'll speak to the general on their behalf. Maybe they see something of interest, because the split is coming. Independence is coming soon. We all know that." The general grunted.

"Now we have the momentum," Heilberg continued as Gabriel translated. "We have to hold Salva accountable. If I could go see Salva with your dad—I would like him to honor his word by signing the papers. I would like to get that paper signed. I would like confirmation. Once we have confirmation of the deal from Salva and the agriculture minister, then nobody can say anything. It's confirmed by the government of South Sudan. I would like Salva to sign, and we would then shut up everybody. Because then it's not just your father, it's the two most powerful men in the country. We'll shut up everybody."

"Okay, we will talk to the Israelis," the general said when Heilberg was done. The promise of mercenaries had captured his attention. There was an uncomfortable pause.

"And we will make an appointment at Salva's house and go and meet him," Gabriel added.

"THE WORLD IS like the universe—ever expanding," Heilberg had told me before we left for South Sudan. "I focus on the pressure points." We'd met one morning at New York's Regency Hotel, on Park Avenue, close to where Heilberg lived with his wife, two sons, and a cockapoo named Cookie Dough. The hotel's power breakfast—$9 cappuccinos, $28 bagels—attracted the Democratic elite, from Al Sharpton to Nancy Pelosi. This was where John Edwards first met Rielle Hunter. It was also where, thirty-two years earlier, Heilberg, son of a coffee trader, native of the Upper East Side, avowed libertarian, had his bar mitzvah. Everyone knew him by name.

His business model, Heilberg explained, was to find those countries about to break into pieces—to identify the future winners of Africa's internecine conflicts and to be standing with them when it was over. South Sudan was his biggest project, but he was also busy befriending Darfuri rebels in London, oil-bunkering militants in Nigeria, and ethnic separatists in Somalia and Ethiopia, looking to cash in on any commodity—petroleum, uranium, whatever—that might come his way in the wake of independence. The strategy had come to him early in his career, when he was a trader at AIG. "I saw the Soviet Union split," he said. "Saw it up close. I realized there was a lot of money to be made in breakups, and I vowed that the next time I'd be on the inside."

Heilberg's methods were always unconventional. In the 1990s, he would fly one month to Moscow with Hank Greenberg in AIG's private jet, then alone another month to Tashkent, the capital of Uzbekistan, where he stayed in hotels so decrepit that he showered with his socks on and he cut gold deals with associates of the dictator Islam Karimov ("a nice guy personally"). He told me he once followed Hans Tietmeyer, the president of Germany's central Bundesbank, into a public bathroom to catch him off guard. "I asked if they were going to lower interest rates or whatever," he said, "just to see if he would tense up or pee on me or something like that." It wasn't until 2002, three years after Heilberg left AIG with enough money to start his own firm, Jarch Capital, and hire talent on four continents, that a friend told him about Sudan: Africa's largest country, rich in oil, minerals, and land, on its way to splitting in two. Heilberg's first contract in South Sudan, in 2003, had nothing to do with food or climate change: It was an oil deal with the Dinka leadership, who he said had wanted bribes he wouldn't pay and had later ignored the contract entirely. The Jarch board was now populated by rival Nuer.

His farmland venture was a more complex play—a kind of double bet on chaos, on national fragmentation plus international food crisis. The 2008 deal had been signed just as the dimensions of the global land rush had come into view, after food prices suddenly spiked worldwide: That spring, soybeans had doubled, corn and wheat tripled, rice quintupled, and the world's grain stockpiles shrank to a two-month supply. The governments of Vietnam, Cambodia, India, and Brazil banned food exports; hungry rioters took to the streets in countries worldwide. The Murray-Darling basin, along with most of its rice and wheat exports, was decimated by drought. In China's grain-growing north, fifty million acres and six million people suffered the worst water shortages in five years. At Costco and Sam's Club, shoppers were limited to a few bags of rice each. "The world needs food," Heilberg told me. "Thomas Malthus talked about the problem of finite land but infinite growth, and he's been wrong to date—we can use technology to push out more food. But what happens when technology isn't coming fast enough? I think people will panic, especially those who have no land to grow on."

The panic was already beginning. By the time we had our tickets for Juba, China was pursuing land deals all over the globe: 3 million acres in the Philippines; more than 2 million in Kazakhstan; 25,000 in Cameroon; 200,000 in Russia; untold tens of thousands in Brazil. Korea, itself faced with water shortages, pursued 670,000 acres in Mongolia, nearly 2 million in Sudan, 3 million in Madagascar—a deal that failed after it helped spark a coup. India, with its population booming and monsoon beginning to shift, pursued 850,000 acres in Ethiopia. More than 1 million in Madagascar. More than 20,000 in Paraguay and Uruguay. Qatar sought 100,000 acres in Kenya. Kuwait, 300,000 in Cambodia. Saudi Arabia, 1.2 million acres in Indonesia, 1.2 million in Tanzania, 1.2 million in Ethiopia, and, in Sudan, the first 25,000 of 250,000 acres of wheat and corn. The United Arab Emirates leased 800,000 acres in Pakistan and as many as 250,000 in Ukraine, 125,000 in Romania, and a million in Sudan. Starting in 2009, thousands of Emirati sheep were landing at the Ho Chi Minh City airport; they were raised by Vietnamese farmers, then slaughtered and shipped back to Abu Dhabi.

"Too bad you weren't here a week ago," Heilberg said. At 7:30 a.m., the Regency's dining room was already filling with men in suits and women in pantsuits. "I was here with Joe Wilson"—Ambassador Joseph Wilson, then the vice president of Jarch's board—"and Sean Penn," who would play Wilson in *Fair Game,* the movie about the Valerie Plame spy scandal. "I thought Sean was a great guy," Heilberg said. "I like guys full of passion. The only problem is that it could snowball. He has a bit of a rebel, wild side to him. I could see myself connecting to that. If you start feeding off each other, it doesn't end until something dangerous happens." A waitress came by. Heilberg ordered a skim latte, an egg-white omelet, and a side of turkey bacon.

"Wall Street used to be a straight line," he said. "It made people a lot of money. A lot of money! But the mundane bores me: so the interest rate on this is 6 percent, and you borrow at 3 percent, you know, and so you'll make . . . anybody can do that. As an entrepreneur, it's never a straight line. When you're an entrepreneur, you have to create something." His favorite author, he said, was Ayn Rand, who, like Heilberg himself, had believed that pursuing profit was itself a moral act, a kind of enlightened selfishness: Place yourself above all else; get in no one's way, and let no one get in yours; give no charity, and expect none. "Her individualism is extreme—but anything in its purest form is more powerful," he told me. "Howard Roark is the hero of *The Fountainhead* because he's pure. He doesn't care about what anyone else thinks—not about social norms, the right clubs, the right people. We all want a bit of Howard Roark in us."

Heilberg was proud to be getting his hands dirty in Sudan. It felt pure. Big banks, Goldman Sachs in particular, would soon be accused of distorting commodities markets, overwhelming the grain exchanges of the American Midwest with speculation—winning paper profits from paper gains while producing nothing tangible. Prices were becoming volatile. "If food stocks are low, a small shortfall in production can cause a big jump in price," explained Nicholas Minot, a senior fellow at the

International Food Policy Research Institute. "Demand for food is inelastic. People will always pay to keep eating." The world could say what it would about him and his generals, but Heilberg knew he was betting not on volatility, not on a bubble, but on something real: an actual food shortage. "We already have a commodities problem," he said. "I would not be surprised if in a day or a week oil goes to $150 a barrel. Like that: Boom! We're seeing the death knell of the financial instrument—of the paper world. We're going to see the rise of the commodity. A bushel of corn can't be $15."

The food crisis could be blamed on a host of factors—climate change, soaring oil prices that jacked up the cost of fertilizer, China's growing appetite for meat, a global population racing toward nine billion—but Heilberg didn't spend much time thinking about which was which. Insofar as he believed in climate change, he believed in the effects, not the causes: the desertification and droughts and fights over water and land that only made his farmland investment smarter.

Partway through our breakfast, he waved the waitress over: "Is there a way to make it a little cooler in here? I'm dying. It's hot, right? Isn't it? It's hot in here." He caught me looking over his shoulder. Al Gore, wearing a black blazer, had sat down at a nearby table. "Oh, Al Gore," Heilberg said dismissively. "He comes here a lot."

IN JUBA, HEILBERG'S HOTEL of choice was a collection of well-appointed shipping containers known as the Sahara Resort. Prefab containers—modular offices, modular housing—were all over the city, popular because they were portable. They had been driven in by mostly foreign entrepreneurs after the end of Sudan's civil war. If fighting reached Juba again, they could be driven back out. Off the paved road, Juba's dirt streets were yellow and rutted, and they were jammed, most days, with SUVs: the Land Cruisers and Pajeros of aid workers, the Humvees of corrupt officials. The SUVs were otherwise parked at the containers, and the containers were ringed with guards and concertina wire, and when night fell and it was cool enough to sit outside, an air of palace intrigue was everywhere.

"Those two men over there, they are spies," Gabriel whispered one evening at the hotel. "Which ones?" Heilberg asked loudly. "Let's go over and say hi." The men were Arabs in slacks and button-up shirts, one with a mustache and one without. They glanced occasionally across the courtyard in our direction. "It's not like I'm trying to hide," Heilberg said. Round, tall, voluble, he was hard to miss. "This is silly. I'm going over there." Gabriel cringed—not in fear, but at the breach in protocol: "No, Philippe . . ." Heilberg stood up and strolled slowly by the spies' table, just for sport, nodding politely at them as he passed.

The 2011 referendum on southern independence, promised by the north-south peace accord, was ever closer on the horizon. But what paralyzed Sudan this week was an imminent decision by a European arbitration court on who—the Arab-led, mostly Muslim northern government in Khartoum or the African-led, mostly Christian government in Juba—could claim the contentious Abyei region, a flash point on the de facto border. The year before our visit, ethnic clashes in Abyei had burned entire towns to the ground, the worst fighting since the civil war, and now both sides were mobilizing again in advance of the decision.

The spies could have been agents of Khartoum, where the president and indicted war criminal, Omar Bashir, feared losing the southern oil fields—some in Abyei, some in neighboring South Kordofan and Matip's home state of Unity—that constituted 95 percent of Sudan's petroleum production and 65 percent of its national budget. Or they could equally have been agents of Egypt, which had spies in Juba because it viewed the 2011 referendum as a threat to its own national security: An independent South Sudan meant an independent straw in the overstretched Nile, more dams, more upstream agriculture. A third of Egypt's population worked on farms, and the country, promised 75 percent of the

Nile's flows under the colonial-era Nile Waters Agreement, already used more water than was being naturally replenished. Before the Arab Spring, Cairo had done an assessment of Egypt's vulnerability to climate change: Even if its population didn't grow, even if there were no new dams, it would run out of water before the century was up.

The farmland grab was in many ways a water grab. Pre-partition Sudan, Africa's largest country and second-largest recipient of annual food aid, after Ethiopia, was also its richest in water. This was notoriously not the case in Darfur or Abyei, where, as the scenario planner Peter Schwartz had pointed out, shifting precipitation fueled battles between Arab herders and African farmers. But the Nile basin itself was filled with streams, dotted with swamps. The government in Khartoum, where the two forks of the Nile meet, had turned over nearly two million acres to Saudi Arabia, Egypt, Jordan, Kuwait, and the United Arab Emirates, hoping to make the northern region the breadbasket of the Arab world. South Sudan was on the White Nile. Up the other fork, the Blue Nile, Ethiopia announced plans to build one of the world's biggest dams, the 6,000-megawatt Grand Ethiopian Renaissance Dam—which it claimed was only for hydropower, not irrigation. Experts suspected otherwise: Ethiopia is, along with Sudan, one of the major targets in the global farmland rush.

What little agriculture that existed in South Sudan was mostly small-scale: families with a few cows, tiny plots planted with sorghum and corn. Heilberg imagined the landscape transformed by American-style agribusiness, complete with irrigation and fertilizers and four-hundred-horsepower combines. Other South Sudanese were coming to him with deals, he told me. The king of the Bari ethnic group wanted to sell him some land. A Nuer commissioner in Upper Nile state wanted to sell him some more. We had loose plans to check out the former by jeep, the latter by helicopter—and, Abyei tensions permitting, we hoped to fly over the million acres in Unity by airplane.

Heilberg planned for Jarch to farm the land itself with joint-venture partners, not flip it, and to sell crops here before selling internationally. There was a local market for it: Sudan was in the midst of a long-running famine, neighboring Kenya had an accelerating drought, and aid groups were willing to pay top dollar for food. He hated the aid groups otherwise—they were bloated, he said, and they corrupted the economy by handing out bribes and favors, propping up the Dinkas—but he would gladly sell food to them. This was business. The joint-venture partners could be Israelis, perhaps. "They have experience in Africa," he said. "They've shown an ability to figure out problems." He liked the idea of bringing in Israelis to farm what some consider Arab land—a way to show his disdain for Bashir. "Do you know what a tefillin is?" he asked. "The box and leather strap you put on during prayers? It's a reminder of God bringing us out of Egypt or whatever. I always bring mine to Sudan."

The land Matip had granted him was mostly empty, Phil assured me—mostly unused by local herders and farmers. He hadn't surveyed it all himself, but he seemed to believe this. A later study by Norwegian People's Aid of twenty-eight foreign and domestic land schemes under way in ten South Sudanese states would claim that Heilberg's million acres were among the most densely populated: 24.3 people per square kilometer. His holdings covered 80 percent of a county containing 120,000 people—too many to easily resettle. Sudan scholars pointed to a worrisome precedent: During the height of the civil war in the 1990s, according to Human Rights Watch and other witnesses, Matip's private militias had brutally cleared civilians from their homes—torching villages, raping women, executing men—to make way for oil drilling.

"There are no white hats here," Heilberg said. "It's the Wild West. People get upset when I say you've got to go to the guns. Hell, you had to carry a gun back then. You were a cowboy? You would lose all your horses and cows, your women would be raped, and everything you had would be gone. People take their ideals and try to impose them someplace else. That's colonization to me. I don't do that. This is what it is. I'm not promoting it or demoting it; I'm just part of the system."

It was the Dinkas and their Western allies who were writing history, bad-mouthing Matip, Heilberg reminded me. And even they respected power. "This is Africa," he said. "The whole place is like one big mafia. The general is like a mafia head. That's the way it works." Lawlessness had a libertarian upside. "My view is that you want government to be as small as possible," he said. "I don't want someone saying, 'Thank you for your investment, now get out,'" he said. "I want a country that's weaker. There's a cost to dealing with strong countries: resource nationalism. People forget that."

One night, Heilberg had Gabriel take us to meet a key Jarch board member, General Peter Gadet, at an outdoor bar on the banks of the Nile. On the way, we drove down a bumpy dirt road lined with dozens of bombed-out tanks, skeletons exposed by the beams of our headlights, their turrets bent, their tracks missing. South Sudan had new tanks now—including the ones Somali pirates had inadvertently captured in 2009 when they hijacked the Ukrainian cargo ship MV *Faina*. They could help win independence, with or without the referendum.

Gadet, a feared tactician who had rejoined the southern army along with Matip, his fellow Nuer and frequent ally, was then in charge of the south's air defenses. He sat alone with two bodyguards at a table near the water.

"How long are you in Juba now?" Phil asked him. "And how is your family in Nairobi? And do you have anti-aircraft? You have the anti-aircraft? What about the ones that have the wings that go higher? You have this now? Okay. Good." Gadet was a devout Christian, Phil had told me. He'd spent nine years in the bush outside Juba before the peace accord, plotting to seize it.

"And the tanks?" Heilberg asked. "Where are the new tanks?"

Gadet pointed across the river toward the bank he'd once prowled during the civil war.

"Right here?" Heilberg exclaimed. He peered into the darkness.

A Canadian soldier stands guard at the edge of the Northwest Passage,
an emerging shipping lane as the Arctic melts.

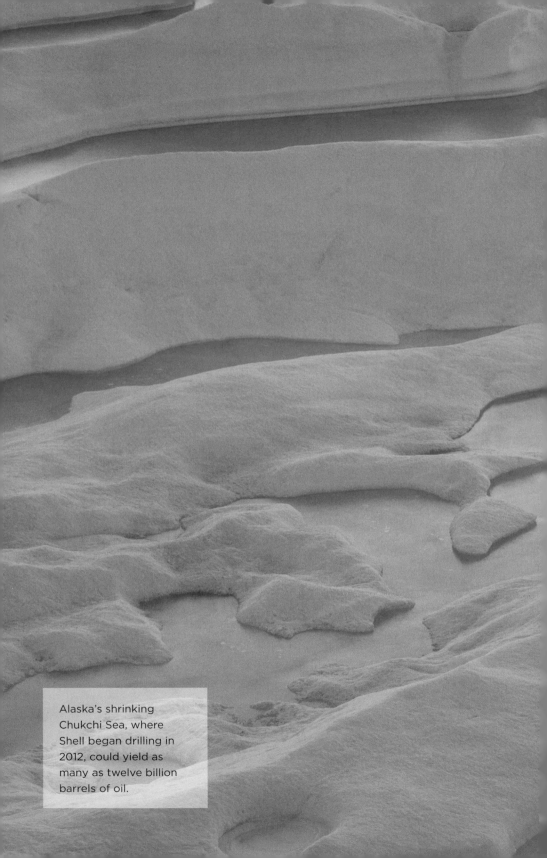

Alaska's shrinking Chukchi Sea, where Shell began drilling in 2012, could yield as many as twelve billion barrels of oil.

Norway's Snøhvit, or Snow White, is the northernmost natural gas facility in the world—and some oil companies' model for the future of the Arctic.

As Greenland's glaciers recede, revealing mineral deposits, mines like Black Angel are expected to help fund its push for independence from Denmark.

Fjords in Greenland are melting earlier and freezing later, extending the season for shipping and icebergs.

On the hunt for potable water, the Israeli desalination engineer Avraham Ophir invented the world's greatest snowmaker, a product now in use in the melting Alps.

Private firefighters working for private insurers race
to protect a client's home in Los Angeles.

During a historic drought in Southern California, the All-American Canal was reengineered so that less water would seep across the border to Mexico. San Diego gets the savings.

Volunteers march in celebration after the first full season of planting trees for Africa's 4,700-mile Great Green Wall, an attempt to block the expanding Sahara.

As food prices spiked, American investor Phil Heilberg (RIGHT)
cut a deal with Gabriel Matip, the son of a South Sudanese general,
for millions of acres of farmland.

Dhaka, Bangladesh, grows by half a million people a year as migrants flee cyclones and rising seas. The world's longest border fence awaits those who continue to India.

The Netherlands, protected from storm surges by massive barriers like the Maeslant, pictured here, is selling its flood-fighting expertise to a worried world.

After crashing into the shore near Kodiak Island, Alaska,
Shell's main Arctic drill ship, the *Kulluk*, awaits repairs in a protected bay.

"How are all your bullet wounds?" he continued. Gadet had been shot twenty-eight times during the war. "You don't need a bulletproof vest. You know who's on your side." He pointed to the heavens. "Always missing the vital areas—that's good. Soon, south will be independent.

"I don't think any war will last long," Heilberg said.

"Short war," Gadet said.

"Short war. I agree."

"Yeah, the war good for them."

IN NEW YORK AND LONDON, I met other investors who, like Heilberg, were stepping away from the paper world. A banker who had told me about Ukrainian farmland-for-vodka deals had me over to his apartment, an airy corner loft in Tribeca, and talked on the condition of anonymity. "Here's the trick," he told me. "All these collective farms collapsed once they decollectivized them, because they had no capital—the guys couldn't afford tractors." This was why vodka and a few months' worth of grain went so far. Via a long-haired middleman he nicknamed Jesus, his investment bank, one of Wall Street's big three, pursued not only thousands of acres of prime farmland but ostrich farms, a chocolate factory, and a Ukrainian pornography channel. The bankers flew through the countryside in a massive, double-rotor Soviet helicopter, landing in fallow fields and peasant villages, and helped introduce a crop of genetically modified, drought-resistant sorghum first developed on an Israeli kibbutz. "You massively raise production and all that," the banker said, "but it was basically a big rip-off of peasants." The Ukraine deal had ultimately fallen through—Jesus had demanded bigger and bigger cuts—but climate change was an area of endless growth. When Europe launched its emissions-trading scheme, doling out carbon permits to coal plants and power utilities, the same banker had helped them "massively overrepresent" their emissions, then helped them sell the excess for hundreds of millions of dollars. "I was actually doing the carbon deals," he said. "All that kind of shit. That was a big scam, too."

The savviest farmland buyers saw global warming as a double boon. In the short term, it was a push factor, fueling droughts that destroyed entire harvests in China, Australia, and the American Midwest and causing food prices to spike. In the longer term, it was creating a pull: Higher-latitude countries like Ukraine, Russia, Romania, Kazakhstan, and Canada are becoming more productive, not less, as the climate heats up. "You don't need to be a rocket scientist to suggest that production belts in the Northern Hemisphere are shifting northwards," said Carl Atkin, the head of agribusiness research at the British real estate behemoth Bidwells, when I met him in London.

One of the many interested firms that called Heilberg after news of his South Sudanese land deal broke, Bidwells had its London offices in a narrow building down a narrow alley off Hanover Square. In a bright fourth-floor conference room, with skylights and hardwood floors, Atkin showed me a world map depicting soil qualities—the USDA's Inherent Land Quality Assessment—with the richest areas shaded in green. "You've got a splotch in North America," he said. "A splotch in South America. Pockets in the U.K. But the main interest is this black soil going up through Russia and the Ukraine: some of the best soil in the world." Environmental factors—frigid winters and short growing seasons—had conspired with political factors to keep prices low; a hectare of black earth in Romania was a fifth of the price of a hectare in England. Overlay a climate-change map on the soil map, Atkin said, maybe add population data, and you could make a fortune. He himself had just returned from Ukraine, and Bidwells had been bringing financial clients to Romania for five years, doing what Atkin called parceling—a plot-by-plot approach to big land purchases. "We reaggregate small plots that were reallocated to everybody post-Communism," he said. "You're getting loads of villagers in a room with the mayor, and the mayor is saying, 'All right, who wants to sell their plot, and who doesn't?'"

As climate change pushed farming to higher latitudes, the money followed. Two of the most visible farmland investors—the British-run Landkom and the Swedish-run Black Earth Farming—had invested hundreds of millions of dollars in agricultural operations in Ukraine and Russia. BlackRock, the world's largest asset manager, invested $250 million in British farmland, France's Pergam Finance sunk $70 million into former ranches in Uruguay and Argentina, and Calgary-based Agcapita put $18 million into Canada's future corn belt. After Saskatchewan land prices jumped 15 percent in 2008—the largest increase on record—Agcapita began raising its next $20 million. But it was One Earth Farms, a sprawling grain and cattle venture on tribal lands in the Prairie Provinces, that would soon be Canada's biggest farm. With funding approaching $100 million and investors ranging from the former prime minister Paul Martin to the agribusiness giant Viterra, one of Deutsche Bank's climate-fund picks, it was forming partnerships with more than forty First Nations tribes who controlled more than two million acres across Alberta, Manitoba, and Saskatchewan.

In Brazil, the British-run Agrifirma, partly owned by Lord Jacob Rothschild and headed by Jim Slater—who came to fame writing an investment column for the *Sunday Telegraph* that he signed "The Capitalist"—had spent $20 million to buy or option 170,000 acres and survey another 6 million. The seed money for Agrifirma had come from the Brits' previous venture, Galahad Gold, which had made 66 percent annual profits by flipping a uranium and molybdenum venture in melting Greenland. Brazilian agriculture would be relatively protected from the ravages of climate, Slater wrote, because the country has "about 15 per cent of the world's recoverable water supply—90 per cent more than its nearest rival."

Banks with prominent climate-change funds, including Deutsche Bank and Schroders, also had separate farmland funds, and in 2011 it was revealed that university endowments, including Harvard's and Vanderbilt's, were invested in London's Emergent Asset Management,

run by the Goldman and J. P. Morgan alumni Susan Payne and David
Murrin. "Climate change means some places in Africa will be drier and
others will be wetter," Murrin told Reuters. "We'll be looking to take
advantage of that." In his 2011 book *Breaking the Code of History,* Murrin prophesied a global-warming-fueled commodities crisis that would
help push a declining West and a rising China into armed conflict.
Africa would be the linchpin, and Emergent's funds, including its African farmland fund and a new climate-change fund focused in part on
water projects, were apparently the best way to profit in the meantime.
By one rough estimate, Emergent had sunk more than $500 million into
agriculture projects everywhere from Mozambique and South Africa to
Zambia.

Heilberg was a bit player in comparison, and his holdings in South
Sudan were not quite as fertile as those in Ukraine, Argentina, or South
Africa: a splotch of blue on Atkin's global soil map, a shade away
from perfect green. But Sudan, already buffered from drought by
the Nile, may be one area that gets wetter: While climate models for
semiarid Africa are notably inconsistent, under some of them Sudan's
precipitation will rise. Warming could be good in general, Heilberg
thought. "Maybe it means we can live in the Arctic," he told me one
morning. "The Nordic countries seem to have a good balance. Maybe
we can have Greenland—there's a lot of land there." On his laptop,
he kept a file on Greenland. It wasn't about farmland. He knew Greenland was rich in minerals, and he'd heard it had a secession movement
of its own.

THREE DAYS PASSED IN JUBA, and it began to look as if nothing would
happen. No new land deals. No overflights or jeep tours. No meetings
with Salva Kiir. Heilberg sat for hours in the air-conditioning at the
Sahara Resort, smoking cigars and playing Texas hold 'em on his Black-
Berry, waiting for word that the president or a minister was ready to see

him. He read the book he'd brought along on the trip: *Sailing from Byzantium: How a Lost Empire Shaped the World.*

"Let's go see Dad," he exclaimed when Gabriel loped into the hotel one morning. Gabriel told us that the Matip family compound in Unity state had just been attacked by a longtime rival, part of the bubbling tension as the south inched toward independence. A guard had been captured and beaten. The general, already stressed about Abyei, was too angry to see anyone, his blood pressure dangerously high. "He is feeling some pain here," Gabriel said, pointing to his stomach.

"When things aren't going well, he internalizes it," Heilberg said softly. "Everybody's looking to him for responsibility. I think it eats at him."

"Okay, so what's the plan for today?" he asked Gabriel. "Have you found anyone for the agriculture?"

"The minister isn't available," said Gabriel. The upcoming Abyei decision was distracting everyone, he apologized.

"Okay, so that's tomorrow," Heilberg said, turning the conversation to new land acquisitions. "So we'll find out about the Bari? What about Upper Nile?"

Soon, there was little more business to discuss. We sat in the lobby and talked about life. Heilberg told us that he'd given up his favorite diet cherry Pepsi—he worried the artificial sweetener would give him Alzheimer's. He told us about the really hot personal yoga instructor he'd had, a point of domestic tension: "I'm in my living room doing my yoga, and my wife flips out!"

Gabriel told us that his own wife, who had cost him eighty-nine cows, had just run out on him. Heilberg offered him some tea. "We Are the World" blared from the hotel's speaker system.

"Did you call Dr. Joseph?" Heilberg asked. The doctor—a powerful figure in Unity state who served as the south's minister of health—was another Jarch board member. "Call Dr. Joseph, and see if he's around."

Dr. Joseph wasn't around.

"We Are the World" came on again—the music was on a loop.

"Is that Bruce Springsteen?" Heilberg asked. We fell silent, listening.

"I think that one is Michael Jackson," said Gabriel.

"Michael Jackson," Heilberg agreed. We waited for the next verse.

"That's Cyndi Lauper," I said.

"Bob Dylan," Heilberg said next.

"What's this guy called, the blind one?" asked Gabriel.

"Uh, Ray Charles . . . no, Stevie Wonder," Heilberg said. We waited. "And there's Ray Charles!"

Gabriel finally secured a meeting with Dr. Joseph for 6:00 that evening. Heilberg put on a dark suit and dark tie; Gabriel wore a gold-colored tracksuit. Just before sunset, we climbed into the Land Cruiser and bounced along potholed dirt tracks in the direction of the jebel, Juba's landmark mountain, passing an open-air market and a field of huts that had just been razed by the government. Now the area was populated by mounds of burning trash.

Dr. Joseph lived in a rare non-shipping-container house surrounded by a thick white wall and attended by a servant, who sat us on plush faux-leather couches below a languid ceiling fan and offered each of us a Coke and a bottle of water. Across from us were three Sudanese dignitaries watching a Nigerian soap opera on a wide-screen TV. "Stay away from my wife!" yelled one actor. "Which wife?" asked the other.

Dr. Joseph had yet to return home. We sank back into the couches. Heilberg began talking nonstop, filling the silence, keeping up appearances. That a supposed ally was too busy to meet with him—that his farmland deal could be speculative in more ways than one—didn't fit the Randian narrative.

He counseled Gabriel about his dad. The general refused to leave his compound until he could discuss the recent attack with the president. "It's called anxiety," Heilberg said sympathetically. "Everyone gets it. It builds up in your mind." Gabriel looked worried.

"Have you ever seen the movie *Analyze This*?" Heilberg asked.

"Analyze what?"

"No, no—it's a movie with Robert De Niro. He's a mob boss, and he's getting really angry, and Billy Crystal tells him, 'You know what I do when I'm really angry? I hit a pillow.' So he takes out his gun and starts shooting a pillow. Crystal's like, 'Feel better?' and De Niro says, 'Yeah, I do.' Your dad needs to feel better. I know the way he is. He should get it all out."

We waited until 10:00 p.m. Dr. Joseph never showed. When we left, Heilberg looked sick. "Are you sure he's still with us?" he asked Gabriel.

But any hit to his confidence was buried, temporary. The next morning, Heilberg was back to his old self. If any of his allies had been bought off by the Dinkas, it was only more proof that Matip and Gadet needed to sweep away the rot.

A week went by. We drove to Matip's compound. We drove back. We drove to Gadet's compound. We drove back. We smoked hookahs at the hotel. We ordered pizza. We drove to the Nile. We drove back. To an outsider in Juba on the cusp of independence, listening to the Nuer whisper about the Dinkas and the Dinkas about the Nuer, waiting for meetings, waiting for savanna to become farmland, waiting for Abyei and sovereignty and all else, it was like being in a hall of mirrors. Either Heilberg had South Sudan all wrapped up, or he had little but his capitalist ideals and a strange friendship with some Nuer generals.

On one of our last mornings, Gabriel disappeared. He didn't show up at the hotel. He stopped answering his phones. Finally, just before dinnertime, he walked in. "These guys were following me on the road," he said. He'd pulled off on a side street, and one of his pursuers had cut around him, blocking his escape. When the man stepped into the road, holding a gun, Gabriel ran him over. The second attacker came up from behind. "I opened the door and hit him, and then he fell on the ground," Gabriel said.

"What did you say to them?" I asked.

"We don't say anything," Gabriel said. "I collect their phones and their guns."

"And kick them in the nuts!" Heilberg said.

By now, Heilberg wasn't bothering to ask about signatures or meetings. The next day, when the European arbitration court announced its ruling on Abyei, it wasn't so bad for the south that it caused immediate fighting, but it wasn't so good that it calmed anyone down. Just in case, the Juba's cell phone service was jammed. There was nothing to do but watch CNN. Heilberg was waiting for his generals to clean house, to create a country where business could get done. He looked at Gabriel. "I was there with your father two years ago, when he told them he would burn down Juba," he said. "I think this is coming. It is coming soon. It'll be short."

We sat with the general once more in the courtyard, guards and elders and wives flitting by in the shadows, a television flickering a few feet away. Matip was still slouching, but this time he looked Heilberg in the eye as he spoke. "What he can tell you is this," Gabriel translated for his father. "All the things going on here, they are not good. You should go. Go to America, and he will call you. He's not happy about the way the government works. He will find out what things happen, and why. In a short time, we'll rise, and put on the Internet, and you will read in America."

"Thank you," Heilberg said. "I know you'll be successful. I agree: The way they're doing things can't last. History has shown us that's how revolutions happen. But I hope you will call me back soon, documents in hand, and we will all smile and be happy. Luckily, it's not all up to us." He gestured to the sky. "There is a higher power."

That night, Heilberg went straight to his shipping container, downed a dose of NyQuil, and passed out. The next morning, we were on a flight back to Nairobi.

It was the same view out the window as before. Green. Heilberg's million acres were in the opposite direction, but the soil was similar—just fewer stones. Nuer tribesmen liked to boast about how fertile the land was. Plant a mango tree, they said, and it would be waist-high in six months. Plant green beans, and the vines would be waist-high in weeks. Plant anything, and it would grow. There might be more war before Heilberg tilled his first seeds. He could wait for it. Demand for food was inelastic.

"Which do you think matters more in Africa?" he asked me. "Military power or political power?" He was sweating in his seat, wearing a baby-blue Lacoste shirt with an alligator on it, and had just been listening to Depeche Mode's "Personal Jesus" on his iPod.

"Military," I said. He nodded. "People say it's going to be north versus south," he said. "I say it's going to be a free-for-all. It's going to be a free-for-all for about a week. Mass hysteria. Juba burned to the ground. Khartoum burned to the ground. Then we'll look around and see who's still standing. They'll form a new government. A period of chaos isn't a bad thing. It'll release that tension. You can't escape the physics."

Like Greenland, South Sudan would soon vote yes in its referendum—an overwhelming 99.57 percent in favor of independence. In the months that followed, northern forces occupied Abyei, burning *tukuls* and hospitals and driving thousands of civilians from their homes, and launched a brutal bombing campaign in the nearby Nuba Mountains. Less noticed outside Sudan was Peter Gadet's post-referendum rebellion in Unity state, fought against the fledgling South Sudanese government in the very fields that Heilberg may someday farm. It was chaos.

"The reason I'm so open with you is so you can see I'm not a bad man," Heilberg told me on the plane. "I'm a guy with a big heart who also wants to make some money." He put his headphones back in. "You know what I give them? I give them hope."

EIGHT

GREEN WALL, BLACK WALL

AFRICA TRIES TO KEEP THE SAHARA AT BAY; EUROPE TRIES TO KEEP AFRICA AT BAY

The main highway out of Dakar, a band of blacktop linking the crowded Senegalese capital and the empty Sahel, was dusty and clogged on a summer day—jammed not only with cars but with people. Young men walked against the flow of traffic hawking peanuts, inflatable airplanes, steering wheel covers, oriental fans, telephone cards, and shrink-wrapped apples. Others stood where the sidewalks would have been, manning makeshift kiosks that sold French-language versions of Yahtzee and Monopoly, posters of sheikhs and imams, and drinking water in plastic sandwich baggies. The highway led to the desert, and the youths of Senegal were doing what they could to go in the opposite direction. Sell enough in a day, and they might be able to afford rice, the national staple, which now cost twice as much as six months earlier. Sell enough in a year, maybe two, maybe five, and they might be able to pay a smuggler to take them to Europe. Every minute or two, a new group approached our jeep, waving their wares expectantly. My host, Colonel Pape Sarr, a thin man who greeted everything else with a cavernous smile, wore a blank expression, staring resolutely ahead into the haze.

I was across the waist of Africa from Heilberg's Sudanese tracts, some three thousand miles west, in the country that imports more food per capita than any other on the continent. Senegal gets three-quarters of its staples from abroad, including 150 pounds of rice a person each year—even as it, too, is a target of foreign farmland buyers. India would soon announce a 370,000-acre deal with Senegal's Ministry of Agriculture, while Saudi Arabia's Foras International would lay claim to 12,000 acres of rice paddies in the fertile Senegal River valley, the first piece of a planned 500,000-acre megafarm. But the scheme that drew Pape Sarr and me to the Sahel, the arid borderland between Africa's humid tropics and the encroaching sands of the Sahara, was something else entirely: the Great Green Wall, Africa's own response to climate change, a forty-seven-hundred-mile-long, ten-mile-wide barrier of trees meant to keep the Sahara at bay. If completed, it would cross eleven countries from the Atlantic Ocean to the Arabian Sea, from Senegal in the west to Djibouti in the east. Pape, a camouflage-clad officer of Eaux et Forêts, Senegal's directorate of water and forests, was one of its architects. We were driving to see his men put the first seedlings into the ground.

The Great Green Wall was proposed in 2005 by Nigeria, where officials claimed desertification was consuming some 900,000 acres a year, and in 2007 it was officially endorsed by the African Union (AU). But in every country except Senegal, it so far existed only on paper. Standing before the press at the Copenhagen climate conference, Senegal's president at the time declared that his nation would be like "the old Greek philosopher" Diogenes, "who proposed that you could prove the existence of motion" by standing up and walking. Senegal wouldn't wait for AU studies or UN approval or World Bank funding. It would prove the viability of the wall by going out and planting—and was hoping that the money would catch up. The government framed the Great Green Wall as a matter of national survival. "Rather than let the desert come to us," said the minister of agriculture, "we will take the fight to it."

It was convenient to think of the Sahara's advance as being like that of a slow-moving army, a single front line against which another easy-to-imagine line—of trees—could form the perfect bulwark. But in the roughly ten billion acres of slowly degrading dryland regions around the world—not only in West Africa, but also in Spain, China, Australia, Mexico, Chile, and nearly sixty other climate-threatened countries, rich and especially poor—desertification is usually a messier process. "The Sahara spreads rather like leprosy," wrote the Briton Wendy Campbell-Purdie, one of the first to try to check the Sahara with large-scale plantations, in her 1967 book, *Woman Against the Desert*. "Little bad spots here and there go unnoticed, until suddenly the whole area is infected."

As a barrier to such an insurgent threat, most scientists agreed, a phalanx of green was largely futile. As a symbol, however—of the protective crouch the world was beginning to adopt in the face of warming, of Africa's particularly lonely position, of how much money rich, high-emissions countries would pay to save themselves from warming's effects versus how little they would pay to save poorer countries—the Great Green Wall was much more potent. For me, it represented a shift toward the third stage of humanity's response to climate change: engineering as refuge, when talk of opportunities rings especially hollow and we begin erecting our defenses. For developing, mostly agrarian countries, closer to nature, this means defenses against what nature is becoming. For richer countries, it means the same thing, plus something more: defenses against migrants and other spillover.

We drove east and then north in Pape's jeep, the land becoming increasingly yellow, the traffic increasingly thin, and began passing billboards celebrating the president's other prestige projects: Plan GOANA, announced after street protests during the food crisis, aimed to expand domestic rice production fivefold by 2015. Plan REVA, or Retour vers l'Agriculture—"Return to Agriculture"—was GOANA's controversial predecessor. Funded in large part by Spain, which in 2006 saw its Canary Islands deluged by more than thirty thousand boat people from Senegal, REVA aimed to turn unemployed youths into agricultural workers rather than illegal migrants. REVA's test cases were planeloads of deportees recently returned from Spain under an arrangement with the Senegalese government. Street-smart young men were promised a hundred hectares and subsidized seeds, expected to remake themselves as farmers, and were so angry at the government's complicity in their forced return that they formed the National Association of Repatriated People. The Great Green Wall was seen in the context of these other make-work projects—especially as it picked up international support. "Failure to act now," wrote the authors of a joint 2009 European Union–African Union study, "could result in many land users becoming environmental migrants—potentially transferring problems north." Whatever else it was, it was in part a scheme to keep Africans out of Europe.

In Touba, a Sahelian boomtown founded by the country's most famous Sufi mystic, Pape stopped to confer with another Eaux et Forêts official. I took a moment to walk amid the minarets and madrassas on the city's dusty streets, where I met boys selling stacks of tapes and CDs to passing cars. "Take me to America," said one. "I will go to Europe," said another.

Pape and I drove onward through the garrison town of Linguère and into the Ferlo, an expanse of featureless savanna named after a long-dry riverbed, where Fulani nomads made their encampments and parrots flitted through sparse trees and yellow grass. The paved road became a dirt road, and the dirt road became a pair of faint tracks, and the jeep began bucking like a horse. At dusk appeared dozens of parallel ditches running through the red earth, some dotted with faint tufts of green. Pape turned proudly toward me. "This," he said, "is the Great Green Wall." The trees were eight inches tall.

ON A MAP, Senegal, almost as close to Brazil as it is to the Spanish mainland, is not an obvious launching point for sub-Saharan Africans trying to reach Europe—not even for reaching the Canary Islands, which sit in the Atlantic due west of Morocco. But GPS technology had turned everyone into a navigator, the Canaries had unpoliced flights to the rest of Spain, and seemingly easier routes—across the Mediterranean via the heavily patrolled Strait of Gibraltar, over the newly heightened fences from Morocco into the Spanish enclaves Ceuta and Melilla—had been successively sealed off. In the months before I reached Senegal, migrants set out daily from the beaches of M'Bour and neighboring fishing towns, each paying nearly $1,000 to be ferried nearly a thousand miles to southernmost Europe. The boats were fishermen's pirogues: brightly painted wooden canoes equipped with two engines and two GPS units and packed with dozens of young men. The passengers' mantra, a mashup of French and local Wolof, was "Barça ou Barzakh"—"Barcelona or Death." Some pirogues capsized in storms; some simply disappeared. The weeklong crossing sometimes stretched to two weeks when captains got lost, leaving boats desperately short of food and water. In record-breaking 2006, thousands died en route—as many as six thousand, according to a Spanish estimate, meaning that one out of every six migrants got death, not Barcelona.

There were idle fishing boats in towns like M'Bour because Senegal was running out of fish. The country was running out of fish in large part because industrial trawlers from France, Spain, Japan, and other foreign countries had been scouring the coast of northwest Africa since at least 1979, when the European Union negotiated its first fishing deals in the region. Over the course of twenty years, Senegal had signed seventeen different agreements with the EU, the most recent one the same week an EU-commissioned study found that the biomass of key fish species had declined by 75 percent in Senegalese waters. Gone were the schools of lucrative tuna, gone were the sharks, and left behind were smaller herring, along with unemployed fishermen, who found new work as human traffickers. In 2009, a University of East Anglia study of the effects of climate change and warming oceans on fishing economies suggested a further problem: Out of 132 nations surveyed, Senegal was the fifth most vulnerable.

Whether the men fleeing for Europe should be considered some of the world's first climate refugees was debatable. If creeping sands and emptying oceans were pushing them out, cities and distant countries, with their promises of electricity, jobs, and education, also exerted a pull. Senegal's greatest population flow was internal—from rural to urban, hut to slum—and it followed a pattern being repeated across the globe in the new millennium, the first time in human history when more people have lived in cities than in the countryside. Rare was the Senegalese migrant who went directly from Sahel to pirogue. Rarer still was one who could point to a single cause—the changing climate—in explaining his move. But the many factors, in aggregate, were exactly what Europe feared. Africa would warm 1.5 times faster than the rest of the world, warned the IPCC (Intergovernmental Panel on Climate Change)—and the Western Sahara region would warm the most. "Climate change is best viewed as a threat multiplier which exacerbates existing trends, tensions and instability," wrote the Spanish diplomat Javier Solana, the EU foreign relations chief and former head of NATO, in 2008. "There will be millions of 'environmental' migrants by 2020 with climate change as one of the major drivers of this phenomenon . . . Europe must expect substantially increased migratory pressure."

Today's boat people could be but a hint of what was to come. And the Continent's response, notwithstanding its efforts at emissions cuts in Copenhagen and at other climate summits, was also a hint of what was to come. It was creating a "Fortress Europe," in the words of Amnesty International—an "armed lifeboat," in the words of the journalist Christian Parenti.

Senegal, Africa's testing grounds for the Great Green Wall, was also Europe's testing grounds for a virtual wall to keep Africans out. The European effort was not as conspicuous as the new fence I saw near the All-American Canal along the United States border with Mexico— where by 2080, according to a recent Princeton study, climate change's effects on agriculture will cause the exodus of up to 10 percent of the adult population. Nor was it as conspicuous as the twenty-one-hundred-mile fence India was completing around sinking Bangladesh or the twin fences Israel announced in 2010 to seal off the Sinai from sub-Saharan migrants. But it was comprehensive: Spanish and Italian patrol boats, emblazoned with the logo of Frontex, the new, pan-European border agency founded in 2005, were already cruising the Senegalese coast by the time I arrived. European planes and helicopters ran aerial surveillance. Soon, a satellite link would connect immigration-control centers in Europe and Africa to help track boat people, and the Continent would be secured by the proposed European Border Surveillance System: a complex of infrared cameras, ground radars, sensors, and aerial drones. The European Parliament would pass its controversial Return Directive, a common deportation policy that allowed migrants to be held without charge for up to eighteen months before being shipped home.

Spain, known for its comparative tolerance of immigration, was trying to offer carrots as well as sticks. It opened six new West African embassies in four years under its migration-focused Plan África, and its development spending jumped sevenfold. Before Spain's recession sent unemployment rates to heights not seen since the dictatorship of Francisco Franco, it began a quota program for guest workers. If they came to Spain legally, laborers escaping high food prices and barren seas could win yearlong stints on massive corporate farms or in a still thriving fishing industry. Some Senegalese got contracts with Acciona, one of the world's biggest builders of desalination plants, which Spain was constructing at a frenzied pace matched only by Israel and Australia, trying to keep up with its own drought and desertification.

Spain spent millions of euros each year luring northern European tourists to its beaches. At the height of the Canaries crisis, it launched a marketing blitz in Senegal, too. With the help of the advertising multinational Ogilvy, it plastered Dakar's buses with images of shipwrecks and ran radio ads warning of the dangers of illegal migration. In one television spot, the legendary Senegalese singer Youssou N'Dour sat alone in a wooden pirogue, waves crashing in the background. "You already know how this story ends," he said in Wolof. "Don't risk your life for nothing. You are the future of Africa."

IN THE FERLO, planting operations for the Great Green Wall were based in a former German research station in the village of Widou Thiengoly, a collection of mud homes and stick fences surrounded by trampled red earth. Next to a dirt soccer pitch was the village's communal well, dug by the French in the 1940s, where nomads with donkey-drawn carts spent hours filling water containers made of plastic or from old truck inner tubes. A makeshift nursery of tree species selected by the Great Green Wall's scientific committee—hundreds of thousands of acacia, balanites, and ziziphus seedlings, their roots wrapped in black plastic bags—was behind the three-room building where Pape and I stayed with the other officers. The building had couches and buzzing flies and ancient electric fans, and in the corners were stacked piles of decaying German pulp fiction, from *Unruhige Nächte* (Restless nights) to *Suche impotenten Mann fürs Leben* (In search of an impotent man). We took our meals here, the Eaux et Forêts men arguing the finer points of planting while eating with their hands from a shared platter. After dinner, we cut the lights so we had enough power to run Widou's only TV. The screen attracted dozens of villagers and hundreds of giant moths, and under the African sky we watched a French-dubbed Jack Bauer fight Middle Eastern terrorists in Los Angeles.

The officers' argument the first morning, held over a loaf of bread and a pot of coffee, was about precipitation. The so-called rainy season in this part of the Sahel lasted but a few summertime weeks, making the planting schedule all-important. If the seedlings went in before the last rain, they might live; if they didn't, they would almost certainly die. Pape, who was forty-eight—he was born two months before Senegal's independence—laid out an idea for a new planting regimen. "Imaginez," he said. "Imaginez!" He continued making his point in Wolof, then turned to me to translate. "The problem here is the rain," he said solemnly. "There is not enough." A tall captain heartily agreed. "C'est vrai!" he exclaimed. Outside, a truck began blaring its horn, and I went to the nursery to watch dozens of men clamber into the back, then cheer as it took off for the front at a wild clip. Nearby, a bucket brigade loaded a second truck with seedlings, carefully passing them one by one, hand to hand, until the bed was full. It rumbled off after the first, kicking up a cloud of dust.

Dirt roads radiated out of Widou like spokes, and after the officers dispersed, Pape and I followed one southeast until there was nothing but savanna grasses and baobabs. After thirty minutes, we passed a cluster of army-green tents—forester housing, Pape said—and soon we crossed the parallel ditches of the Great Green Wall, lines in the earth stretching as far as the eye could see. An Eaux et Forêts crew, a hundred or so machete-carrying youths in jungle fatigues, was waiting next to a water truck. Different parcels of the wall were being planted by different groups—from Eaux et Forêts to local villagers to members of the foresters' union to university kids recruited by the Ministry of Youth—and Pape liked to foster friendly competition between them. It was a contest measured in seedlings used and hectares planted, and naturally he believed that his own men and women were the fastest.

This was a new parcel—a tractor had just dug the trenches—and the workers asked us to plant the first trees. One handed me a seedling and lopped off the bottom of the plastic wrapping with his machete. I took off the rest of the plastic and lowered it into the hole; a few inches of rich, wet soil were now all that insulated the roots from the cracked, sandy Sahel. Seven paces away, Pape planted the second tree and sprayed it with a few drops of water. Before we drove off, he gathered his crew for a speech, imploring them never to tire. "Fatigue?" he yelled. "Non!" they responded. "Fatigue?" "Non!" "Fatigue?" "Non!" "Fatigue?" "Non!"

Morale was important because, as I soon learned, money was scarce. Every time Pape needed to pay for more seedlings or repair a truck, he went begging to the director of Eaux et Forêts, and the director went to a minister, and then they waited. "Wait. Wait. Wait. We don't know how he gets the money," Pape said. There was support for the wall from Europe as early as 2009, but the money—a little over $1 million—went only to a feasibility study. In 2011, the UN's Global Environment Facility (GEF) made headlines by pledging up to $119 million to build the Great Green Wall—but this was not additional funding, it clarified. If the eleven countries involved wanted to starve other projects and divert all the money, this would be allowed. And the GEF suggested that the name Great Green Wall should be used to brand a raft of development projects in the Sahel—dams, wells, animal husbandry—and not, in fact, to build a wall of trees. "In my vision of the Great Green Wall, there will be practically no place for plantations," the GEF's Senegal program officer told me. Even if Western money did get diverted to Pape's trenches, the EU would be spending at least ten times more on a virtual wall around itself than on a green wall around the Sahara.

So far in Senegal, international support for the Great Green Wall came mostly from the Japanese spiritual group Sukyo Mahikari. In Japan, the Mahikaris' main temple is an architectural marvel: five minarets topped with five Stars of David surrounding a cavernous hall with a traditional Japanese roof, inside of which is a koi-filled aquarium and a wall of water spewing from the heads of Mayan gods. In the Ferlo, they camped in green military tents in the savanna forty-five minutes from Widou. They had giant bonfires and religious lessons when not planting, and they marched around camp, chanting.

"Have you heard of them?" asked one of the Eaux et Forêts lieutenants in the jeep one afternoon. I'd looked them up, I said: "They believe in the healing power of light energy." The lieutenant nodded. He was sweating and had bloodshot eyes, a likely case of malaria. "Un sect," he said—a cult. "All they do is pray."

Beside us was a civilian Eaux et Forêts official named Mara, a man of feline movements who described his job as "évaluation": He traveled around the agency's various projects, taking notes and asking long, philosophical questions. He had been staring out the window at the empty trenches of the Great Green Wall. "It's good to believe in something," he said. "It helps you do things."

WHEN FRONTEX INTERCEPTED the flat-bottomed pirogues close to the Senegalese coast, the European Parliament's resident migration expert told me, the encounters were unpredictable, sometimes violent. The men were at the beginning of their journey, and they wanted to go on. "They can be ferocious," he said. "Sometimes, there is even the throwing of machetes. But if we approach them out in the open sea, when they have been out for a good amount of time, they are too exhausted to offer any resistance." He cocked his head. "This is interesting."

Simon Busuttil was slight and soft-spoken, with a young, clean-shaven face but graying hair. From Malta, one of the EU's newest members and certainly its smallest, he was the 122-square-mile island nation's senior representative to greater Europe. I met him in his parliamentary office in Brussels. When I was in the Ferlo, Busuttil had also been in Senegal, riding on Frontex boats and meeting with top ministers, trying to determine if Spain's cooperation agreements with Senegal could be a model for all of Europe. Before the toppling of Moammar Gadhafi, he had made a similar trip to Libya, and after the Arab Spring sent waves of refugees across the Mediterranean, he would lead a delegation to Tunisia. "In fact, I was in Washington last September," he said.

"The U.S. Coast Guard gave a very interesting presentation. Between the Dominican Republic and Puerto Rico, the influx has been reduced from around four thousand a year to one thousand. They use biometric equipment. We have some lessons there to learn.

"Finding countries that cooperate is not so easy for the EU," Busuttil said. "So in that respect, Senegal is a blessing." But from the perspective of some in Malta, the Spanish-led clampdown was also a curse. As the Atlantic route to the Canaries was blocked, the flow of African migrants was shifting: across the Sahara they went, on trucks and on foot, from Mali to Niger and then into Libya, where smugglers packed them into matching wooden boats—black hulls, black decks, black gunwales— and sent them in the night across the Mediterranean. In 2008, after the clampdown began, there was a 70 percent drop in migrant arrivals in the Canaries. That year in Lampedusa, the Italian island near Malta in the central Mediterranean that in 2013 would host Pope Francis's first official trip out of Rome, there was a 75 percent jump in arrivals. More than thirty-one thousand people washed up, almost the exact number that had reached the Canaries at their peak. "If you close a door, people will try to go through a window," Busuttil said. "This is human nature."

Data on Mediterranean casualties was incomplete—fishermen who caught bodies in their nets sometimes threw them back, so onerous was the paperwork if one found a dead African—but one in twenty-five boat people was said to die in the crossing. Malta was the next-worst fate: a wrong turn on the way to neighboring Lampedusa, which migrants targeted because they could expect to eventually be transferred to the Italian mainland. Malta, on the other hand, was no ticket to the rest of Europe. When illegal migrants landed here, usually by accident, they were thrown in jail for up to eighteen months. When they were released, they had nowhere to go, because Malta is so small. It is the eighth most densely populated country in the world, just ahead of Bangladesh. Arrivals to Malta were few in comparison to Lampedusa—only twenty-seven hundred in 2008—but in Maltese terms this was another twenty-two people per square mile, and its nationalists were becoming enraged. Under existing EU law, the country where migrants first land is responsible for them. If they escaped to another European country and were caught, they would be transferred back to Malta.

"All evidence points to them not wanting to come to my country," Busuttil emphasized. "They're trying to go to Lampedusa. Those who manage to get through our defenses, they are thinking they've reached Italy. When they see the flags and realize they haven't, they get the shock of their lives." In the Parliament in Brussels, he vainly pushed for what he called "burden sharing": a recognition that frontline states such as Spain, Italy, Malta, and Greece were now policing the border for all of Europe. The Continent's richer, more northerly countries, great emitters of carbon and producers of wealth, barely contributed ships or aircraft to Frontex, and they processed a relative trickle of African asylum seekers. So Malta was itself a victim in this, he suggested. It was a power game: Northern Europe bullied southern Europe. Southern Europe fought within itself and with or against North Africa. The big stepped on the small, and the small stepped on the smaller. The migrants themselves were at the bottom. Here, too, shit rolls downhill.

When vessels were detected drifting between Malta and Italy, the two countries sometimes fought over who should take them in. In 2007, twenty-seven Africans were left clinging to a tuna net because the boat's Maltese captain refused to bring them on board; they had to be rescued by the Italian navy. Italy later pulled out of the Frontex missions altogether and cut a quiet side deal with Gadhafi. "What will be the reaction of white and Christian Europeans faced with this influx of starving and ignorant Africans?" the dictator told the press on a state visit to Rome, as he angled for $5 billion a year to stop smugglers' boats from leaving Libya. "We don't know if Europe will remain an advanced and united continent or if it will be destroyed as happened with the barbarian invasions." Under the 2009 Treaty on Friendship, Partnership, and Cooperation, Italy had agreed to pay 250 million euros a year for twenty-five years in exchange for joint infrastructure projects, oil contracts, and help on immigration. Rumors circulated of vast detention camps in the Libyan desert, and the migrant flow seemed to shift farther east. The Greek-Turkish border became Frontex's next hot spot.

Malta's up-to-eighteen-month imprisonment for new arrivals was its own deterrent, Busuttil told me. "Not because we want to be cruel to other human beings," he said, "but because it's our only remaining weapon, if you like. It might seem harsh, but if you knew the context of this country . . ."

North of Malta, a web of detention centers was rising all over Europe: more than two hundred sites spread between two dozen countries, from a former Jewish internment camp in France to an abandoned tobacco factory in Greece to an empty airline hangar in Austria. Together, they had room for as many as forty thousand migrants. In Britain, most of the prisons were run by private contractors such as the Serco Group, the MITIE Group, and especially G4S, the world's second-largest private employer, after Walmart, and the source of a scandal at the 2012 London Olympics: Its guards were so poorly trained that the British army had to be called in to replace them. The contractors carried out deportations, too, escorting shackled Nigerians or Angolans or Bangladeshis on a coach-class flight home. Outside the EU, G4S ran Australia's refugee detention system until another scandal, this one involving children sewing their lips shut during hunger strikes. In the United States, the prisoner market was dominated by the Corrections Corporation of America, whose lobbyists helped draft Arizona's controversial 2010 immigration bill, apparently because it was good for the bottom line: The more migrants arrested under the new law, the more demand there would be for the corporation's jails.

Climate change would only grow the market, as I was reminded after I left Busuttil and drove across the border to the Netherlands, a partly below-sea-level country preparing for migrants as well as a more literal deluge from sea-level rise. North of Amsterdam, amid shipyards and warehouses in the city of Zaandam, was a new, 544-migrant prison: two gray modernist blocks ringed with concertina wire and built not on land but on water. The prison floated. Until a guard chased me away, screaming, for whatever reason, in angry, Dutch-accented German, I paced along its fence line, snapping photographs of water lapping up against the cell blocks.

I'D TRIED TO ARRANGE for a translator in the Ferlo, and one day, after Pape, Mara, and I returned to the research station from the Great Green Wall, he was there. Magueye Mungune was twenty, a village boy in hip-hop garb: white baseball cap cocked sideways, baggy jeans, expensive shoes. His home was forty miles south across the plain. There he was alone among old people, he told me. All his friends had long ago left, some for Dakar, some for other cities, some for Europe. His mother was in Mauritania, his elder brother in New York. "But if everyone leaves, who will stay for Senegal?" he asked. "People will do jobs in Europe that they would never do here. I do not want to clean toilets!" Magueye had heard rumors about the Great Green Wall, heard that the government was planting trees to stop the Sahara from swallowing his and the other villages, and he already knew what he thought about it. He thought it was stupid.

"They will never finish it," he whispered. "There is no water. You need people to water the trees, but soon there will be no one here. The trees will die. The ministers will just eat the money." Pape, lounging in a nearby chair, took the newcomer in. "If I were a minister, I would do the same," he joked.

"No, no, we must set an example," one of the officers said. "We should take the money for the Green Wall and just give it directly to the people. And forget the wall." Magueye, unsure if the foresters were making fun of him, was momentarily disarmed. "How long will it take to plant?" he asked. Pape did the math: It was almost seven thousand kilometers across Africa to Djibouti. The wall would be fifteen kilometers wide. So, 10.5 million hectares of trees. This summer, his teams would plant 5,000 hectares. "Okay, it will take twenty-five or fifty years," Pape said. "Just for Senegal."

Magueye deciphered the good-natured arguments that raged end-lessly around me: Mara asked about cows eating the seedlings, and Pape said they would not be interested in acacia or ziziphus. The experts on the scientific committee had thought of this already. Mara asked about goats, and Pape said they would eat the trees if they had a chance—but they would not have a chance. The goats belonged to the nomads, and the nomads left the Ferlo at the end of the rainy season, which was before the Eaux et Forêts men would leave. As for the people themselves, they would not cut down the Great Green Wall for firewood because acacia trees were worth more alive than dead: They produce gum arabic, a hardened sap that is used in everything from marshmallows to M&M's to shoe polish. To make one point, Pape pulled out the *Dic-tionnaire de l'écologie* from his briefcase, a black leather sack with an ivory latch, and waved it in Mara's face. Sometimes he stopped to con-sult a weathered printout of Senegal's forestry code or to leaf through a manila folder marked "GMV"—for Grande Muraille Verte, the Great Green Wall—which contained the minutes of scientific committee meetings. Sometimes he sounded like a true believer. "This is like extending electricity across Senegal," he told Mara. "In 1967, the coun-try said, 'Let's do it.' And in 1968, we did it."

The Great Green Wall had inherited its name from the Great Wall of China, the barrier that repelled invaders for two thousand years. When it failed, it was due to human weakness, not any engineering faults: A corrupt seventeenth-century general had accepted a bribe and let the Manchu army sweep past. More immediately, Africa's wall had been inspired by China's own Green Wall, a planned 2,800-mile bulwark against the drifting sands and dust storms of the Gobi Desert. Its first poplars and eucalyptus trees were planted thirty-five years ago. It was already the largest man-made forest in the world. There were other precedents: Joseph Stalin's Great Plan for the Transformation of Nature involved planting interconnected bands of forest across the southern steppes. In the Dust Bowl–era United States, the Great Plains Shelterbelt Project planted 220 million trees stretching 18,600 miles, from North Dakota to Texas. In Australia, in addition to the Number 1 Rabbit-Proof Fence of the early twentieth century, which wasn't entirely rabbit-proof, there was MOTT: Men of the Trees, a nonprofit that since 1979 has planted 11 million seedlings. MOTT was the inspiration for the *Woman Against the Desert* author Wendy Campbell-Purdie's 1959 move to North Africa. She grew trees twelve feet high in the Moroccan Sahara, then moved to Algeria, where her plantation in a 260-acre trash dump led to a countrywide project. First known as the Green Wall, in 1978 it became the Green Belt of Northern African Countries—and then, with loss of interest and the passage of time, again became the Sahara.

Even in Senegal there were precedents, smaller scale tree-planting programs to the west and south. "But those ones are not the Great Green Wall," reasoned Mara. "*This* is the Great Green Wall." Magueye and I looked on as he and Pape argued about why this was important, and soon they looked back at us.

"What is your philosophy?" Mara asked Magueye. "What is mankind's proper relationship with nature?" My translator, sitting in one of the compound's few chairs, sank further behind his white baseball cap. "Do you like Kant?" Pape asked. "Descartes?" Magueye said nothing. "What is spirituality?" Pape continued. "What does it translate to?" Finally, Magueye pointed to the sky. "There is something looking down on me," he said. "But I can't say who."

These questions soon seemed relevant. One afternoon Pape received a situation report, and an air of crisis settled over the research station. The professionals of Eaux et Forêts were no longer winning the friendly planting competition. The light-energized spiritualists of Sukyo Mahikari were faster. "Sukyo Mahikari?" someone asked incredulously. "Sukyo Mahikari?" Pape seemed resolved. "We must go faster," he said. "We must be the fastest."

We piled into a jeep to see the Mahikaris in action. Their camp was orderly—green army tents in military rows surrounding a main pavilion of blue tarps, with a kitchen area to the side—and everyone wore a white name tag. The president of Sukyo Mahikari Senegal, whom Pape and the others called Monsieur Président, stood in the shade of a baobab and held court with two of the prettiest volunteers, his hands clasped together, his head giving small, formal nods: Japanese in manner but otherwise a six-foot-tall African man in a gray tracksuit. A videographer, who also had a name tag, filmed the scene. Two other leaders, a Frenchman from Toulouse and a South African woman from Cape Town, spoke to each other in Japanese, and when they called the camp to attention, the youths—Congolese, French, Senegalese, Ivorians, Guineans, Gabonese, South Africans, Belgians—yelled in response: "Hai!"

Out in the Mahikari parcel, pickup trucks delivered seedlings to donkey carts the group had managed to hire, which dispersed them further to pairs of boys with cloth slings, who were loaded up until their backs sagged and they lurched forward to the planting crews. "Arigato!" the planters cried when they arrived. Pape and I walked along beside them, gaping at the efficient supply lines, sweating in the sun. One woman, a leader, followed a few paces behind the advancing army. She stopped at each new seedling, aimed her open palm at its tiny leaves, and nodded beatifically as she beamed in invisible bursts of light energy.

The jeep was silent on the way back to Widou, the officers still in shock that foreign amateurs—cultists!—could be better than their own. But after a few minutes it began to rain, and our windshield darkened as nearly a season's worth of water began falling upon the dry Ferlo. The ground was so hard that it couldn't soak it up, and everywhere puddles grew. The greater good came back into focus. "It is raining," Pape proclaimed in English, and he smiled at the sky.

It was carnival when I landed in Malta, and the island nation's European population packed the streets of its walled capital, Valletta, parading and drinking before the onset of Lent. Confetti littered the cobblestoned streets, and fluorescent floats—cartoonish red hearts, giant white horses, cardboard kings, an apparent mashup of an octopus and the Statue of Liberty—careened past Baroque architecture, flanked by dance troupes. The dancers wore painted masks and angel wings and colors as bright as the floats, and some of the women had powdered their faces pure white. On the main tourist drag, all the shops were shuttered save for an appliance store. Nearby, I saw the first Africans. They stood in the shadows, clutching their new purchases—electric hot plates, phone chargers—and waited for the procession to pass.

Outside Europe, Malta is known mostly for its Knights: a Catholic military order founded in Jerusalem around the time of the First Crusade, then transplanted here, where they repulsed the Ottoman Empire's three-month siege in 1565. Six years later, Malta's powerful fleet formed the backbone of the Holy League, which destroyed the Ottomans' naval power in the Battle of Lepanto. The Mediterranean was becoming a Muslim sea before the Knights had their victory, and thereafter it was Catholic, Christian. Tiny Malta still commands a search-and-rescue zone far out of proportion to its size, which partly explains why its military intercepts so many migrant boats, so many of which happen to be filled with Muslims. Today, the Knights are headquartered in Rome, a chivalry with sovereignty but no territory, and Malta is traditional, homogeneous, and still Catholic, a near monoculture inhabiting an island almost as dense as Singapore.

The reality of Malta's population density—3,360 people per square mile, about one person every fifty feet—hit me when I got lost on a

Sunday drive, stuck on a series of ever narrower country roads, but never once lost sight of hikers, farmers, or other cars. I ended up near the airport. The sun was setting, and at every pullout, every patch of gravel big enough to fit a Fiat, men and families had parked their cars. They were packed bumper to bumper, yet the doors were mostly closed, the windows mostly rolled up. Nobody talked or looked at one another, though some people ate fast food. Most were just staring into the distance—enjoying, or trying to enjoy, a private moment.

Most of the island's African population was also near the airport, in a cluster of jails and tent camps. I returned to the area one morning to meet with the commander of air, land, and sea operations for the Armed Forces of Malta (AFM). Lieutenant Colonel Emmanuel Mallia was younger than I expected, with slicked-back black hair and a widow's peak. He wore horn-rimmed glasses that gave him an intellectual air and sat at a wooden desk, where he read off statistics from the screen of a silver Acer computer. It was only February, but the year had already seen 530 arrivals—four boatloads. Big boats. The previous year, 2,775 migrants came, but in eighty-four boats. Small boats. Most of the time, they were spotted before they landed. "It's practically impossible for a boat to come here and not be seen by someone," Colonel Mallia said. "If you think you are alone in Malta, think again." Often, they were towed in by the AFM. The military had a force of seventeen hundred people and a maritime budget of almost $10 million, and among its many duties—defense, presidential protection, airport security—border control was now prime. Intercepting and managing boat people had become 80 percent of its work. Every time the members of the military approached a migrant vessel, it was a negotiation. "If they say they want to go on, they can go on," Colonel Mallia said. "If they say, 'We are lost. We need information,' we give them the information they require. When it is bad weather, they want to be saved, but when it is not, they get picky. So if they don't want a rescue, we give them life jackets. That's a lot of life jackets, but it's the least of our problems." Whether life jackets emboldened some to push on to Italy in rickety ships, he didn't speculate. "In summertime water temperatures, you can survive a few hours if you fall in," he told me. "No more than ten. If you are fit and thin, you are the first to die. But if you are fat . . ." He paused. "Actually, we don't get fat ones."

Of those who did land in Malta, 90 percent applied for asylum. The migrants who won it—about half of the new arrivals, especially those fleeing fighting in Somalia or Sudan—also won early release from detention and access to schools, health care, and other social services. But the rest were stuck. There is no asylum for those fleeing mere economic chaos, let alone environmental chaos. Under international law, there is still officially no such thing as an environmental or climate refugee. Migrants denied asylum when I was in Malta before the Arab Spring—Tunisians, Egyptians, Malians, Nigerians, Senegalese—went straight from eighteen months in jail to "open center" tent camps: half-way houses with canvas walls. They were free to come and go as they pleased—but not, of course, to leave Malta, not according to the EU. The open centers I saw looked rather like Sukyo Mahikari's camp in the Ferlo, only surrounded by factories, and they seemed oddly permanent.

Among native Maltese, xenophobia was bubbling up in the street and on Internet message boards. The island's fear of being overrun had led to arson attacks on Jesuit Refugee Service (JRS), which helps the migrants with asylum applications. JRS's director, Joseph Cassar, told me how someone had set fire to their young lawyer's front door, then her car. Other cars were scratched with keys and had their tires slashed, and inside the nonprofit's compound, which was itself once attacked with Molotov cocktails, six vehicles were burned "until there was nothing left, just the metal shell." A short-lived political party, Azzjoni Nazzjonali—National Action—was trying to channel the fear into electoral power by pledging to clean Malta of "dirt, corruption, and migrants." The party's co-founder Josie Muscat, a doctor who ran private clinics in Hungary and Libya, complained about the taxpayer money the migrants were wasting. "If they break something in the detention center, they should have to live with it," he told me. "If we're giving them food and water, they should do some kind of work—like fixing roads. I say we keep them in there and don't let them out until they say they're ready to go home."

Before leaving Malta, I had a tour of one of its "closed centers," where migrants did time for their accidental arrival on the island. The Safi Barracks were inside a military base, and after I cleared security, I found myself staring at a field filled with dandelions and more than a hundred haphazardly stacked boats. Cookie-cutter wooden vessels perhaps twenty feet long and scarcely seaworthy enough for three passengers, let alone thirty, they were the previous year's model from the human-smuggling syndicates of Libya. A soldier led me deeper into the base, to where the men he called "clandestini" were held.

The two-story barracks, the site of periodic riots, had a new ring of chain-link fence, and I walked inside flanked by four guards. The biggest among them soon stopped to admonish the Africans in broken English: Using power strips, they had plugged seven hot plates and the floor's only television set into a single electrical outlet. "You will burn the wire!" he yelled, but then the power shorted out anyway, and another soldier walked off to find the fuse box. At one end of the open-air hallway was a makeshift Ping-Pong table: a piece of plywood resting on a trash can, with a net made of a long strip of cardboard propped up by two milk cartons. Only the paddles and ball were real. When the TV came back on, it was playing Maury Povich, and a dozen men stood in the cold February air to watch, crossing their arms and hunching over to stay warm. Detainees often fought over television channels, I was told, but not today, and when I interrupted the show to ask where they were from, they proudly recited their home countries. Ivory Coast. Ghana. Nigeria. Mali. Guinea. There were no Somalis or Darfuris here—only West Africans, whose chances of winning asylum were slim. There were eighty-two men on this floor, and there were four rooms, which were kept warm mostly by body heat and the thin, military-issue wool blankets they hung over the windows. Inside the rooms were bunk beds, and some of the men had pulled mattresses onto the floor so they could sit and play cards or checkers, the latter using dried orange peels and a piece of cardboard. I was embarrassed to be reminded of my own first trips to Europe: The rooms looked and smelled like a youth hostel.

When my French failed and the Ivorians drifted off, two Nigerians, Tony and Kelvin, became my guides. Tony told me he had traveled through Libya, where robberies and random beatings made it dangerous to be a black man out in public. (It would become more dangerous during the revolution against Gadhafi, when sub-Saharan Africans were assumed to all be mercenaries.) Tony was a mechanic and had spent 5,000 mostly borrowed dollars to go to Italy—only he'd ended up here instead. "We stay in detention for one year, and after we get no document," he said. "Do you understand? No travel document. One year. Three hundred sixty-five days make a year. Do you understand?" He led me into his dorm, where graffiti covered a wall: "Jesus, have mercy, oh Lord" and "How long you keep us here?" The authorities provided a rough soccer pitch out back, Tony said, plus calling cards for his floor's communal phone: 5 euros to every migrant every three months. The detention was survivable, just pointless. When it was over, he would be released on the streets, not sent back to Nigeria. He would just be older. "Everyone here, he has something," he said. "Maybe he is a student, maybe a technician. I am a mechanic. Say I stay here one year. I lose some skills—I have to train three to four months just to be ready again. One year. Do you understand?"

Tony and Kelvin and a growing crowd of men led me to the bathrooms, where they pointed out broken stalls and showerheads. "See, this is our bath," Kelvin said. "Cold water. It is not good. It is not fair." Another man tapped me on the shoulder. "Ten months," he said. "Ten months I am here." Another walked me to the sinks. "No hot water," he said. I turned on a faucet and let it run for a minute, then I tested it with my hand. To be fair to Malta, it was lukewarm.

The guards and I were walking out when one of the Ivorians stopped us. He was older than most of the others, a muscular man in his mid-thirties, and he asked me to look at his arms. "I was very strong before," he said. "I want to work—not sleep. I do not want to sleep." I looked at his arms. His voice rose, but he didn't yell. "I work," he said. "I work! I want to work."

WHEN THE RAIN STOPPED in the Ferlo, it was covered with peach fuzz—green grass that briefly made everything seem alive. Local village youths and a crew from Senegal's foresters' union, who had often faced off in soccer matches in the red dirt and goat shit outside the research station, were suddenly combined to plant what Pape called la Grande Parcelle: the summer's largest swath of Great Green Wall at 4,950 acres. Whether it was the rain or the competition with the Mahikaris, the leaders seemed energized. Pape mapped out supply lines. Mara, watching the workers sweat one sweltering afternoon in government-provided T-shirts—"We plant trees to fight the desert"—had an epiphany. "We should not give them shirts," he declared. "We should give them hats."

The Mahikaris finished their parcel a day and a half early: 125 children planted 1,468 acres in five days. "Motivation—that's what they have," Mara said. "Discipline," said Pape. He and a lieutenant were asked to give speeches at the group's closing ceremony, which they gladly did, thanking Monsieur Président for bringing the first true international aid to the Great Green Wall. "One day you will be able to say to your children: I helped build the Great Green Wall!" said the lieutenant. "Thank you for your rigor, your courage, your discipline, your sacrifice," yelled Pape. "You are valiant people. This is one part of a wall that will cross all of Africa. This is the first step!" The Mahikaris, who were standing in formation in their Sunday best—blue jackets, red kerchiefs, white skirts for the ladies, white pants for the men, and white tennis shoes for all—then broke into song about "a wall to Djibouti . . . ici au Sénégal." They paraded around camp, goose-stepping past a crowd of Fulani villagers who had come for the send-off. We stayed long enough to watch them take down the tents. When a metal crossbeam fell on one girl's head, her friends tried to put a compress on the resulting goose egg. One of the leaders pushed them away. With the friends looking anxiously on, the woman held her palm a few inches from the girl's forehead, and in beamed more light energy.

The Great Green Wall might have been futile against the Sahara, but that did not stop Pape from wanting to believe it could be done. Increasingly, I found myself wanting to believe it, too. When the trucks left early one morning for the Grande Parcelle, overloaded with trees and workers, I decided to ride along in one of the water tankers. Combined, the villagers and the foresters' union youths formed a hundred-meter-wide army when we reached the parcel, and they attacked fifteen trenches at a time: diggers up front, followed by tree carriers, followed by the largest contingent, the planters. They marched forward in their sandals and disintegrating tennis shoes, advancing so quickly across the savanna that I had to jog to keep up. A green Toyota pickup with a bed full of seedlings drove between the trenches, and when it stopped, the tree carriers swarmed it, then dispersed: expand, contract, expand, contract, like a jellyfish.

The Toyota was out of trees after little more than an hour, and we waited as the sun rose higher in the sky. Most of another hour passed before there was a resupply, and after twenty minutes we ran out again. The officers grimaced. We found refuge in a cluster of grown trees, fifteen groups in fifteen patches of shade, and the planters, who carried razor blades to remove the seedlings' plastic wrappings, now clenched them in their teeth. They took turns examining my hiking boots, rubbing the leather and tapping at the Vibram soles. We had nothing to drink until the tanker truck arrived, nothing to do until the next load of trees arrived after a two-hour break. It was noon. It was hot. The hotter it got, the slower we went, and it was no one's fault: the Grande Parcelle was too far from the nursery for an easy resupply, and there weren't enough trucks, because there wasn't enough money, because there weren't enough people who cared to see the Great Green Wall become reality.

Pape soon appeared in his jeep, and he still seemed hopeful. When the tanker truck started up again, preparing to head deeper into the parcel, one of the boys, barely a teenager, rushed over to get a final drink. He grabbed a big red plastic cup and filled it to the brim, but he got only one sip before Pape yelled over at him. "Eh, eh," said the colonel, and he pointed at a newly planted row of Great Green Wall. The boy didn't protest. He dumped the rest of the water on an acacia seedling, silently watching it pool around the base of the tree and sink into the earth.

PART THREE

THE DELUGE

A blueprint for disaster in any society is when the elite are capable of insulating themselves.

—*Jared Diamond*

NINE

GREAT WALL OF INDIA

WHAT TO DO ABOUT THE
BANGLADESH PROBLEM

E namul Hoque was an Assamese lord, albeit a minor one, and not long before we met, half his family's land had been washed away when the Brahmaputra River changed course. In his lifetime, he had already been forced to move his home five times. He was thirty-seven. He had a mustache and a slight overbite, and at night, when he lit candles to deal with the frequent power outages, he showed a fondness for whiskey and cigarettes. He was soon to marry a beautiful young girl, a Muslim like him. Until then, he rented a small house near the law college in the northeastern Indian city of Dhubri, where he had a bathroom filled with enormous spiders, a servant with whom he proudly spoke the indigenous Goalpariya language, and a set of maps showing border emplacements, fences, boundary roads, and guard outposts. His life was dedicated to sealing off Dhubri and the rest of his home state of Assam from the people everyone here called "infiltrators": Bangladeshis who snuck across the border for economic opportunity or to escape their own country's host of natural and social disasters, including cyclones, overpopulation, seasonal famines, and especially rising waters that ate away at land and crops.

By the time it crashes into India's restive northeast, the Brahmaputra has cascaded almost seventeen thousand feet from its Himalayan head-waters. From Dhubri, it has just a hundred more feet to lose before reaching sea level—but it has to snake four hundred miles through neighboring Bangladesh before actually reaching the sea. This cannot be done quickly or directly. The formerly steep, clear river is at Dhubri flatter and broader and browner than ever before, and it carries more sediment than almost any other on the planet. It is five miles wide, and it is constantly jumping its banks. Considering its former location, Enamul's ancestral land was surely carried downstream across the border, where it might well have become part of a new riverine island and been claimed by Bangladeshi farmers who had lost their own land—an irony he chose not to focus on.

Enamul had lately become chief of the International Border Affairs Committee for the powerful All-Assam Students Union (AASU), which for three decades had campaigned to save ethnically distinct Assam from what it called "a silent invasion" of Bangladeshis. India, after a big push from AASU, was quietly constructing a $1.2 billion fence around Bangladesh, and Enamul spent his days driving alongside it and taking boats along it and walking it with binoculars, looking for gaps. "I ask, 'What are the lacunas?'" he told me. "What are the plans? And what is the real picture?" When he saw infiltrators, he reported them. When he saw problems with fence construction, he reported them. Once, he walked so many days on the border's sand and loose dirt that his left knee swelled up dramatically. "Like this," he said, and he cradled an imaginary basketball in his hands. Another time, he heard there was a firefight between Indian and Bangladeshi border posts, so he rushed to the Indian side, borrowed a gun from a dead guard, and began shooting. Often, he bragged, he was in such remote border areas that he had to skip his lunch. He was a patriot. He was like one of the activists from America's Minuteman Project, only he was fond of yoga.

At more than 2,100 miles, the new border fence—flanked by new roads, illuminated with floodlights, soon to be electrified—would be the longest in the world. It would be so long not because Bangladesh, with its 164 million people, is large—it's smaller than Iowa—but because Bangladesh is surrounded: The predominantly Muslim country, which in 1947 broke off together with Pakistan from predominantly Hindu India, remains encircled by India on three sides. (Bangladesh's only other land border, 120 miles fronting Burma, is in the process of getting its own barbed-wire fence, and its south is bounded by the ever expanding Bay of Bengal.) When Bangladeshis sneak west into the Indian state of West Bengal, where the people are ethnically and linguistically indistinguishable, they blend in. When they sneak north across a much shorter section of border with Assam, the locals notice an influx of darker-skinned people who speak a different language. This was one reason that the first and loudest calls to seal the border had come from here.

What Enamul wanted was a perfect fence, something that could keep the Bangladeshis out no matter how unlivable Bangladesh became. He had been a communications student when he first joined AASU, but like many people today he now approached social problems with the mentality of an engineer. The question was not what we could do but what we could build, and India's razor-wire-and-steel response to migration— much blunter than Europe's varied responses to its African migrants— seemed to me even more representative of what was beginning to happen in this third stage of climate distortion, as the world faced up to rising seas in addition to melt and drought. Walls. From here on, in one sense or another, this is what those of us who could afford them were engineering against climate change. Those who could not afford them would be stuck on the other side.

India was a poor country, but Bangladesh was poorer. India emitted more carbon than Bangladesh, and perversely this signaled that it had more resources to deal with the effects. The first Bangladeshis had not come to Assam because of global warming, and AASU had not been worried about warming in the 1980s, when it first pushed for the fence.

But it was worried now. "Global warming, if it happens, what will happen?" its leader had asked me. "Will there be war? Will Assam become part of Greater Bangladesh? Most of Bangladesh will be underwater, and where will they come?"

As the fence went up, it was the job of India's paramilitary Border Security Force (BSF) to hold the line. Nearly a thousand people have been shot dead at the border since 2000—about one every four days. In a 2010 report, *Trigger Happy,* Human Rights Watch detailed a pattern of extrajudicial killings and torture: boys killed while fishing too close to the fence, men shot in the back as they tried to run away, criminals armed with sticks felled by border guards armed with rifles. Indian authorities claimed the border's lawlessness—ethnic insurgent groups, smuggling of narcotics and rice, and especially the rustling of tens of thousands of cows that lost their holy status upon leaving Hindu India— justified any violence. In one widely publicized incident, a fifteen-year-old Bangladeshi girl named Felani was shot when she tried to cross back from India, where she was living illegally, into Bangladesh, where she was about to be married. Her purple *shalwar kameez* caught on the barbed wire, and for five hours her dead body hung upside down. "We fire at criminals who violate the border norms," the BSF's director general said during an official visit to Bangladesh. "The deaths have occurred in Indian territory and mostly during night, so how can they be innocent? We have made it clear that we have objection to the word 'killing,' as it suggests that we are intentionally killing people."

I tried to get the BSF's permission to visit the border well before flying to Assam. In Delhi, I called the force's headquarters again and again until an officer relented, telling me I could expect to get my clearance when I showed up in Guwahati, Assam's largest city. In sprawling Guwahati, I took a taxi to the local BSF camp, where they told me they could do nothing without Delhi's okay in writing. In Shillong, in Assam's neighboring state of Meghalaya, I secured a meeting with a deputy commandant, but when I got there after a three-hour jeep ride, he had been called away to a meeting. I finally took an overnight taxi ride toward Enamul's hometown of Dhubri, and by dawn, as we passed through villages along the braided Brahmaputra, it already looked like Bangladesh: streets impossibly packed with cars and rickshaws and pedestrians. Dhubri District had one of the highest population densities in India: 1,492 people per square mile, about half that of Bangladesh. As I neared Dhubri city, I saw another BSF camp, and I decided to try to bluff my way in. I dropped the Shillong officer's name, and a young soldier led me down a long corridor and into a sparsely furnished office, where he made a series of calls while I peered at a document on his desk titled "Unnatural Death." He finally put down the phone and turned to me. "I'm sorry," he said, "but the border area is closed to foreigners." That was the point.

I was in the city scarcely twelve hours before the police showed up at my hotel. "Actually, we are protecting you," said a man in a leather jacket, and I was led down the street to the station for questioning that became all the more polite once they were certain I was American. On the second floor, in a hall with a languid ceiling fan and teal green walls, an officer gently thumbed through my passport while others stared at an ancient television that was showing the movie *Titanic*. One wall had a hand-drawn crime map: recent petty thefts, cattle rustling, and banditry by armed dacoits. Bangladesh was ten miles away. The relaxed mood changed only after I was allowed to leave and the next interviewee was ushered in. She was a small woman in a beautiful orange *shalwar*, towing a young son. "Bangladesh?" the officer asked. She nodded. His smile faded.

LIKE ALL ELSE related to climate change, sea-level rise is not the same across the globe—not uniform, certainly not equal. An extra inch in the North Sea does not necessarily translate to an extra inch in the South China Sea or Sea of Cortez or Bay of Bengal. Satellite measurements cited in the IPCC's 2007 report show two parts of two oceans—the western Pacific and the eastern Indian—rising more quickly than any others, while measurements taken along the lengthy Indian coastline show that some areas, including West Bengal, adjacent to Bangladesh, are more quickly losing ground. The variability is attributed to tectonic movements, to changes in the distribution of heat and salt, which lead to changes in water circulation, and to the fact that surface winds can literally move oceans. The Hadley cell and another atmospheric circulation thought to be invigorated by climate change, the Walker cell, are pushing water from the Indian Ocean's southern reaches north toward the Bangladeshi coast, according to a recent University of Colorado study. And there is another factor causing uneven sea-level rise, the subject of a flurry of recent research, that bodes especially ill for Bangladesh and many other places in the low-lying tropics: The thick ice sheets atop Greenland and Antarctica have a strong gravitational pull on surrounding waters, yet that force is diminished the more the ice sheets lose mass. More melt, less Greenland. Less Greenland, less gravitational pull. The perverse result of Greenland spilling at least fifty trillion gallons of water into the sea each year may be "a smaller sea level rise in the far North Atlantic," explains John Church, the lead author of the sea-level chapter in the IPCC's 2014 report. "Of course, a smaller rise in one place means a larger rise elsewhere."

On average, global sea levels are rising at a rate of about three millimeters a year—twice the rate of the middle part of the last century but still mostly manageable, equivalent to adding little more than an inch every ten years. If the expansion remains linear, oceans will be roughly a foot higher in 2100. But few scientists believe it will stay linear. The summer I traveled with Minik around Greenland, the eight-nation Arctic Council began one of the most authoritative surveys yet of the island's precipitous melt. The researchers found that the flow rate of Greenland's largest glaciers had increased two- or threefold and that small earthquakes—the rumbles of calving glaciers as icebergs fell into the sea—were several times more frequent than in the early 1990s. Thermal expansion, the fact that when water heats up, it expands, is no longer the biggest contributor to sea-level rise, their report claimed. Instead, it's melting ice. An average rise of three feet by 2100 is now considered a reasonable forecast; some experts believe six feet is within the range of possibility.

In the Bay of Bengal, the creep of the sea was like the migrants sneaking into India: silent, mostly invisible, just beginning. Even at six or eight millimeters a year—local scientists' rough estimate—it was having an effect. In the Sundarbans, the world's largest mangrove forest and the last bastion of wild Bengal tigers, high tides and higher salinity were starting to kill the namesake sundari trees, which lost life, leaves, and color from the top down—"top-dying disease," the locals called it. In the adjacent delta formed by the Brahmaputra and the two other great rivers of Bangladesh, the Meghna and the Ganges (known here as the Padma), seawater was not inundating the land so much as infecting it: There is a point in an estuary system where an inflowing river becomes subsumed by the sea, where freshwater becomes so intermixed with salt water that it can no longer be considered fresh. This point was moving inland year after year. Salt levels in the waterways of six southern districts have risen by 45 percent since 1948. The amount of damaged cropland increased from less than four million acres in 1973 to more than six million in 1997 to a projected eight million or more acres today. An Indian dam on the Ganges, the Farakka Barrage, completed in 1975 to divert freshwater to Kolkata, was blamed for worsening the problem—less freshwater down, more salt water up—and now more megadams, Chinese as well as Indian, were planned for the Brahmaputra. Bangladesh was being hit from both sides. In a country where small farmers make up half the population, fields and rice paddies that fed thousands, even millions of people were gradually becoming too salty to sustain crops.

Salinity is one of the four horsemen of climate change for southern Bangladesh described by the country's leading environmentalist, the

IPCC author Atiq Rahman. The next is cyclones. While warming's effect on tropical storms is hotly debated, a general consensus is emerging: Whether or not it increases their frequency, it very likely increases their strength. Cyclones and hurricanes are fueled by ocean temperatures; more heat means more destructive winds. Bangladesh, long in the path of storms, is likely now in the path of larger storms. In late 2007, category 4 Cyclone Sidr—the second-biggest storm since reliable record keeping began in 1877—sliced into the Sundarbans and southwestern delta, destroying 1.5 million homes and killing more than three thousand people. In 2009, smaller Cyclone Aila left at least half a million people homeless. It sent a twenty-foot-high wave crashing over fields and into mangroves: With sea levels higher, there is more water to push around, and storm surges are all the more damaging.

The third horseman is increased flooding. Seasonal flooding is normal in Bangladesh, and in many ways positive. A typical monsoon season sees as much as 30 percent of the country underwater as the Brahmaputra, Padma, Meghna, and dozens of other rivers swell with rain and overtop their banks. Farms flood, families are displaced, and some riverine islands, known as *chars,* disappear entirely. But new *chars* are created as the rivers discharge their billion-plus tons of sediment, and a new layer of mineral-rich soil remains when the waters recede. The soil allows Bangladeshi farmers to plant and sow a remarkable three crops a year. Flooding is like the greenhouse effect itself: It makes life possible. Only in excess—a modified monsoon, a higher Bay of Bengal— does it extinguish it. Flat, slow rivers, having less elevation to drop than before, were becoming yet flatter and slower, and seasonal floods were starting to last longer and spread farther. A Dutch-designed, foreign-financed system of dikes and embankments built in the 1960s was worse than useless: Rather than keeping water out of the delta's cropland, the barriers often trapped water on the wrong side, turning fields into ponds. The Dutch were now hawking updated technologies via their embassy.

The fourth horseman requires little explanation: A sea-level rise of three feet by 2100—or whatever the global average will translate to in Bangladesh—will permanently submerge at least the southern fifth of the country, simple as that. The people who live there, twenty to thirty million of them, will have to go somewhere else.

THERE WAS A STANDARD tour route for foreign journalists reporting on Bangladesh's woes: south on the river ferries from Dhaka, its capital, to the saline, overcrowded delta, to the *chars*, to the Sundarbans, and to the villages flattened by Cyclone Sidr. Almost as soon as I crossed over from India, I found myself following it. My guide for the tour was Atiqul Islam Chowdhury, or Atique, an unflappable, unfailingly polite man in his thirties from the local nonprofit COAST, which focuses on "survival strategies for coastal poor." Our deal was straightforward: I paid his way for what amounted to a week of site visits and meetings that his organization could not otherwise easily afford, and in return he gave me access to villages and to his sense of quiet indignation.

I'd come across the border on the newly reopened Maitree Express, or Friendship Express, a Kolkata-Dhaka train that had been closed for forty-three years and was now touted as the start of a new entente between India and Bangladesh. Much of the 230-mile rail trip was on elevated tracks above the latter's rivers and *chars*, which were visible out the picture windows but blurred by speed and distance, as if part of a different reality. But when Atique and I rode the *Parabat*, or "Pigeon," a large, southbound night ferry from Dhaka, such distance was harder to attain. He had booked us an air-conditioned cabin with a small TV and an outlet to charge my cell phone, but the waters of Dhaka's Buriganga River were only three decks down, the masses only two decks down. There was a blackout in Old Dhaka the night we left, and on the opposite bank was a shipyard where the sparks of welders' torches periodically brightened the humid sky, like lightning. We got chai and samosas from an attendant, then the ferry's foghorn blared. As we pulled away from the dock, we could see the faint hulk of the massive city fading away, and for the rest of the night there was only whatever appeared in the beam of the ferry's spotlight. It swiveled back and forth, operated by rope and pulley. A bearded man stood by, tugging on one side, then the other. The rules of the river were like the rules of the Bangladeshi road: When the spotlight-wallah's beam caught smaller ships or wooden, gondola-like sampans ahead of us, the ferry began blasting its horn, and they got out of our way before we ran them over.

We arrived in the city of Barisal at dawn, and Atique hailed a rickshaw that would take us to the crowded bus that would take us farther south on a series of crowded, two-lane roads. There were huts and rice paddies and palm trees and people and shrimp farms on both sides of the road, but as in Malta not once was there an empty, uncultivated, perfectly natural space. The shrimp farms consisted of rectangular ponds hemmed in by dirt walls, and they were a relatively recent addition to the landscape, the result of two trends: Bangladesh's growing need for export dollars and the delta's creeping salinity. Shrimp has become the country's second-largest source of foreign income, after textiles, and each year as many as a hundred million pounds of it now flow out to some of the world's biggest emitters of carbon: half to Europe, a third to the United States, and most of the rest to Russia and the Middle East. When Bangladeshis gave up farming rice in favor of shrimp, it was sometimes hailed as climate adaptation, but it was lopsided: Shrimp farming requires far fewer workers, and the profits are largely kept by exporters and middlemen. Small farmers did not switch to shrimp themselves so often as they sold or leased their land to the country's small oligopoly of shrimping families, then migrated to Dhaka or beyond.

Where there were still rice paddies, yields per acre were going down. Seasonal food shortages were already such a way of life that they had a name: the *monga*. The Bangladesh Rice Research Institute (BRRI) and international partners would soon begin large-scale tests of salt-tolerant rice, hoping to keep up with the rising seas. The BRRI's first varieties were conventionally bred and distributed for free in southern Bangladesh, though it was also working on genetic modification. Other varieties being developed across the world—including by Monsanto, which had an office in Dhaka despite being dropped from a deal with the country's Grameen Bank after activists protested in the late 1990s—were genetically modified for expected profit. Bangladesh, where the average person emits 0.3 tons of carbon a year—a seventieth of the average American—could do little more about climate change than try to adapt to it. Elsewhere in the country, NGOs were creating floating gardens: mats of water hyacinth covered with soil, cow manure, and seeds that grew into gourds or okra. They turned a riverboat into a floating school, and they bought other boats to serve as rescue arks during flooding, even training locals as their pilots. It was reminiscent of another program in a similar climate-threatened, rice-growing, shrimp-exporting delta, the Mekong in Vietnam, where the government and international donors began distributing life jackets and teaching children how to swim. But only in Bangladesh were families told to raise ducks, not chickens. Ducks float.

Atique and I saw a more typical adaptation effort in the town of Mirzakalu, on the banks of the lower Meghna, the megariver formed after the Padma, Brahmaputra, and hundreds of other tributaries flow together. South of a small ferry dock, a crowd of workers was lifting sandbags, mixing cement, and sliding large stone blocks into place to form a new seawall. It stretched as far down the shore as we could see. The ground was so covered with sandbags that Atique and I skipped from one to the next, as if they were stepping-stones, in order to reach the foreman. "This embankment is temporary," Atique said, even before we got to the man. "Within six months, it will go away. Just look behind it." I looked, and I saw the beginnings of a second, higher seawall five hundred feet inland—the backup plan. The workers gathered around us, and we asked them how many times walls had been built here. They argued among themselves. "Seven, eight times," Atique finally translated. Shortly before Cyclone Sidr, the shoreline had been a mile "that way," one man explained, and he pointed toward the middle of the broad river. The foreman told us that they had been working on this latest seawall for three weeks, laying some ten thousand blocks and forty-five thousand sandbags. The blocks weighed 120 kilograms, or 265 pounds, the sandbags 160 kilograms. Most of the workers came from Rangpur, in far northern Bangladesh, adjacent to Enamul Hoque's hometown of Dhubri in Assam; only Rangpuris could easily lift the blocks and sandbags. Two of the men ripped off their shirts to show me where they carried the loads on their backs. There were but a few scratches and old scars. They found a day laborer from Mirzakalu and had him take off his shirt, too. His back was bleeding from half a dozen cuts.

Some of the locals displaced by erosion had gone to new *chars* still growing in the middle of the Meghna—the sediment of the Himalaya or Assam, deposited here. To visit them, Atique and I climbed into a wooden fishing boat and sailed across the chocolaty water. Soon we were skirting mud flats and entering a canal flanked by low fields, thatched huts, and a few tin-roofed shacks. Children swam in the canal, and fishing boats were moored along its mud banks. The *char,* first settled in the 1970s, was named Zahiruddin. In 2002, the last time it was surveyed, it covered almost fifteen square miles and had eight thousand residents; by now, the population was surely bigger, but no one could say if that was true of the *char* itself. It had no dikes. It had few cyclone shelters—concrete structures on concrete stilts—though the government was building hundreds of them all over the mainland. Its residents were among the most vulnerable people in the world. We docked the boat and wandered around the *char*'s fields, trying to find someone to talk to, but the day was too hot; almost everyone was inside his hut. At last an old man appeared, shirtless and holding a large black umbrella for shade. He showed us the thousands of red peppers he had spread out to dry on three squares of cloth. "He is one of the fortunate ones," Atique translated. "He came from the mainland eighteen years ago, so he has a title to his land." Months later, when Cyclone Aila struck Bangladesh with seventy-five-mile-an-hour winds, it pushed a storm surge far up the Meghna. "Char Zahiruddin completely gone under water," read one NGO's report.

"Want to go to the hotel and get fresh?" Atique asked me. He meant "freshen up." We made our way back to solid ground, and in the evening, after the air had cooled, we went to an outdoor play where performers dressed in green danced on a stage ringed by cloth and held up by bamboo struts. The subject was cyclones, and the play, underwritten by aid groups, was like an extended public service announcement. "It's about what went wrong during Sidr," Atique explained, "and then how the whole family can change and be ready." The crowd of hundreds of men, women, and children sat in the dirt, in the dark, while the cast, illuminated by dim fluorescent lights, played flutes, banged on drums, sang, and yelled. One actor unrolled a large scroll, upon which were painted revolving images of a disaster: A family watching television, and the numbers 1 through 5—the cyclone intensity scale—flashing on the screen. Other people in small boats, huddled around the radio. Rickshaw drivers yelling warnings to those passing on the road. Families grabbing their gold jewelry and anything else portable and rushing out of their homes. Eventually, people filing calmly into cyclone shelters— an image that looked uncannily like the Edward Hicks painting *Noah's Ark*. It was a hopeful play: Early warning systems, cyclone shelters, and education were why relatively few people—3,000—had died in Cyclone Sidr. Sixteen years earlier, in 1991, a similar cyclone had killed 138,000.

A million tons of rice were lost during Sidr, however, and the following spring, in the midst of the 2008 global food crisis, Dhaka was yet another place where riots broke out over the spiraling price of rice: Garment workers went on strike, smashing cars and throwing bricks at police, who responded with bullets.

One morning, Atique and I visited the fringes of the Sundarbans, hopping on motorcycle taxis that sped us down elevated paths to the village of South Khali. This and surrounding villages near the city of Bagerhat—which, like Touba, near Senegal's Great Green Wall, was founded by a Sufi saint—were among the worst hit by the cyclone. Of the fifty South Khali families that survived the storm surge, clinging to palm trees or clustering on the second floor of the combined school–storm shelter, half had since moved away. It was the emptiest place I would visit in Bangladesh. We walked to where the path ran into the Bay of Bengal and watched fishermen untangle their nets. At the yellow two-story school that had saved people's lives, a villager pointed out how it could be improved: If there were an entrance on the second floor, people could get inside even after the flooding began. I noticed the words on Atique's gray T-shirt: "Beach Tour." A fisherman eventually offered to take us across a creek into the Sundarbans, where a forest watchman in a lonely cabin told us that Bengal tigers, their ecosystem disrupted, were killing more villagers than ever before. We ventured a few hundred feet into the brush until we were advised to turn around.

On the ferry back north to Dhaka, Atique was sullen. "In fifty years, all the islands will be gone," he finally said. "It will cause strife. Environmental refugees will have no place to go, or they will go to cities. Talk about Islam, about fundamentalism—these people will be angry. This will cause a war. Americans want to have their houses, their cars. They don't see what it is doing to Bangladesh." He urged me to have a look at the lower decks, so I did. They were more crowded than they had been on our way south. Each family had staked out an area of floor with a blanket, fathers, mothers, and children, together holding down their private patch. Many had large bags—seemingly all their possessions. "Do you know why?" Atique asked. "They are all moving to Dhaka."

IN FACT, SOME AMERICANS—those in the defense establishment—did see what was being done to Bangladesh. In the years following Cyclone

Sidr, Bangladesh was front and center in a series of war games and intelligence reports. One of the largest such games was held in July 2008 at the Center for a New American Security in Washington, D.C. John Podesta, soon to be the head of Obama's transition team, played the role of the UN secretary-general. The featured speaker was the Shell alumnus Peter Schwartz, who had just been running a closed-door scenario involving submarines and Arctic melt for an unnamed client. A separate study he had done on the future of maritime navigation in the high north had recently been featured on *The Colbert Report.* "For the first time, my seventeen-year-old son knew about what I was doing," Schwartz joked at the podium. "[Colbert] made me a hero. But just as important, even Colbert is recognizing the threat of the melting Arctic." He turned more serious. "We are already seeing signs of climate change," he told the war gamers. "This is not a fifty-year issue. It's not, in my view, even a twenty-year issue. It's a today issue, whether it is flooding in Bangladesh, storms in Myanmar, or droughts in Australia."

The script of the war game imagined water tensions between Mexico and the United States, an influx of refugees from the Sahel and North Africa to Europe, the construction of floodgates to protect New York City and Shanghai, mass crop failure followed by mass flooding in India, a category 5 cyclone that killed 200,000 people in Bangladesh, and 250,000 climate migrants camped out at the India-Bangladesh border. The game's outcome, unlike the subsequent, actual events in Copenhagen, was a robust global climate treaty.

The following winter, the National Defense University ran its own version, describing what could happen if millions of flooded-out Bangladeshis streamed into India: Food and water shortages. Epidemics. Religious war. A 2010 exercise at the Naval War College determined that the U.S. Navy would be hard-pressed to respond to a major disaster in Bangladesh without mobile desalination plants and shipboard capacity for thousands upon thousands of flood victims.

The most robust look at the implications of climate change for

Bangladesh, South Asia, and, by extension, the U.S. military was the classified work of the National Intelligence Council (NIC). After its initial, global climate-security analysis, the NIC had gathered more specific climate data for six countries and regions. Bangladesh appeared to be among them. "We take that data," an official told me, "and give it to a group of political and social scientists—people who understand how humans react—and say if this is what happens, given the other things that are going on in the region, how will the people react? We never look at climate change by itself. I mean, you gotta look at it in the context of other issues. Will we see cooperation to solve problems? Will there be tension? Will there be migration? If they migrate, where will they move?"

The reports produced for the NIC by outside defense contractors gave clues that the intelligence community worried about the same things everyone else did. "Anticipated inundation and salt water intrusion in the Ganges delta may displace tens of millions more Bangladeshis," read one prepared by Centra Technology and Scitor Corporation. "India would not have the resources to cope with Bangladeshi immigrants pushing into West Bengal, Orissa, and the Northeast . . . About half of Bangladesh's population, unable to sustain themselves through agriculture, will migrate to cities by 2050, and most of this migration will probably be to India. In addition, major disruptive events such as cyclones may generate mass refugee movements into India on much shorter timescales than the overall shifts in climate."

In India itself, such studies were far fewer. When I visited, there had been only one government report probing the links between Indian security and Bangladeshi demographics, climate change, and sea-level rise—and it was classified. "Apart from its high population growth rate, it's very clear that Bangladesh is going to lose a very significant portion of its landmass," the report's author told me in Delhi. "It is a ticking time bomb."

But India, which was climbing in the ranks of the world's worst carbon emitters—per country, not per capita—was waking up. Forty percent of its GDP is dependent on rainfall, and precipitation, including the timing of the monsoon, was changing. NASA satellites showed groundwater levels in its north falling by as much as a foot a year as irrigation sucked aquifers dry. India was ranked the world's twenty-eighth most vulnerable country on the Climate Change Vulnerability Index put out by the British risk consultancy Maplecroft—well below Bangladesh, which was second, but well above most of the rest. At climate-security conferences, a former commander in India's air force, A. K. Singh, began warning of fights with Pakistan and Bangladesh if shrunken glaciers forced India to keep water from shared river systems on its side of the border. A flooded Bangladesh would further destabilize the subcontinent. "It will initially be people fighting for food and shelter," he told NPR. "When the migration starts, every state would want to stop the migrations from happening. Eventually, it would have to become a military conflict. Which other means do you have to resolve your border issues?"

ENAMUL HOQUE did not dislike foreigners per se, and with me he showed himself to be nearly as gracious a host as Atique. Almost as soon as I had arrived in Dhubri, he came to welcome me at my hotel, which consisted of a few rooms above a clothing shop that were lately being overrun by a hatch of thumb-size grasshoppers. The hotel was "bad for his status," he said, so we drove to his house in a white Tata hatchback he had borrowed from a friend. Along the way, he assured me that I was in good company. "I am a very popular man in Assam, all over India," he said. "But I get phone calls. People threaten me—call me black sheep. I say back to them in Arabic: 'The only God is the homeland. Do everything for your homeland. Otherwise, you are not a true Muslim.' That is what I say to them." The fact that the infiltrators and their apologists were usually Muslim and that he was also Muslim was not as important as the fact that he was Assamese and he was Indian and the infiltrators usually were not.

Inside his two-room house, he had his servant grab a roll of border maps for me, which we spread out on a desk next to the book *You Can Win,* by the motivational speaker Shiv Khera. ("Winners don't do different things. They do things differently.") On a wall was a *jaapi,* traditional Assamese headgear that looked like a mix between a Mexican sombrero and a conical Vietnamese *nón lá.* Enamul sat at the desk and began telling the story of immigration to Assam. The first wave entered as refugees in 1971, when Indian forces helped Bangladesh, then known as East Pakistan, gain independence from West Pakistan. (For two decades after the partition of India, the two were a single Muslim state.) By 1979, AASU's leaders were so worried by the population boom that they launched what is known as the Assam Movement. The anti-immigrant campaign included mass student rallies, sit-ins, and the massacre of 2,191 illegals in one six-hour period. (Enamul didn't mention this.) It so roiled the northeast that the government sat down with AASU and signed an accord in 1985, promising foremost to build the fence. "That is why the work started in 1987," Enamul said, "and that is why out of all the states, the first part of the fence was in Assam."

The maps Enamul showed me were living documents—constantly changing not just because new roads and fencing were still going in but because completed roads and fencing were sometimes consumed by erosion and because new areas of undefended land sometimes appeared in the middle of the river. "See here," he said, pointing. "Once upon the time, miles of border fence were built here. Now it is all dismantled, so we have to rebuild." He flipped to the next map. "This area is a very nice area for crossing," he said. "It is very vulnerable." He pointed to Bangladesh. "Chaos from global warming is starting from that side," he said. "After ten years or twenty years, the Bangladeshi people, they are bound to migrate. Because after ten years or twenty years, Bangladesh is not fit for living for a human being. The situation is very alarming. Presently,

they are coming by any means, and they are searching for their liveli-
hood, and they are settling hither and thither. It is a silent invasion."

Enamul did not dare take me on one of his border patrols without
permission from the BSF, but he had an equally important tour in mind:
He wanted to show me Dhubri as he knew it, what was at stake if it was
overrun by Bangladeshis. The next day, he picked me up on his moped
at 5:00 a.m., and we puttered past shuttered storefronts until we reached
the Open Air Theatre Cultural Complex, where he regularly practiced
yoga with his Hindu friends—more proof that his fight was not sectar-
ian. Dozens of middle-aged people were sitting in auspicious pose in the
dirt, women on the left, men on the right, legs crossed, fingers forming
mudras, while three leaders chanted from an elevated stage. Enamul
and I put down mats on the men's side, and after an hour of my flailing
while everyone stared at me, he was gracious enough not to comment on
my flexibility—once declared "the worst I've ever seen" by a Manhattan
physical therapist. His friend with the car soon came for us, and we
drove out to a famous potters' colony on a riverside spit of land, where
we stood for hours under the beating sun, observing folk artists and
their terra-cotta rhinos and elephants.

The tour's most important stop was in the afternoon: a visit to a
decaying wooden mansion that had been perched on one of Dhubri's
only hills for the last hundred years and to the blue-eyed raja who still
lived in it. To meet him, we parked at the base of the hill and walked
through uncut grass past an old cannon, then up a rickety staircase past
his drying laundry, then into a second-floor study, where he sat surrounded by hundreds of photographs and oil paintings of his ancestors.
For a few minutes, fans kept the room cool, and after the power went
out, the soft-spoken raja offered us cans of Coca-Cola, which he somehow served with ice. "You should not drink the water here," he warned
me. He briefly showed us the armory downstairs, which had dusty
trunks, mounted tiger heads, and an elephant gun, but mostly we sat in
the sweltering study and talked about the old days. "We had seven

hundred square miles," he said. The system of lords and vassals had ended barely two decades earlier, after the government started collecting taxes directly rather than using the raja and lesser nobles like Enamul. The raja's land, which had stretched across Assam and even into present-day Bangladesh, was gradually claimed by the government and distributed to the masses. This mansion was one of the few holdings left. He wanted to turn it into a museum.

The raja's father, a member of the Assamese parliament, had been a great hunter. "He killed seventy-six tigers and eleven leopards," he said, "and he caught more than a hundred elephants." He pulled out an old hunting log, becoming increasingly nostalgic as we flipped through the ragged pages. "One tiger, two male rhinos, and one female," he murmured. "From this, I think we can also make a list of extinct animals." Over time, he said, his father noticed that the tigers were disappearing, and he became a great conservationist. The raja hovered over a photograph of his father's favorite elephant, Pratap, who died in 1962. The animal's grave was in the front yard, near the cannon. "My father was always telling me: Pratap is different. Pratap is unique. Every elephant has a breeding period, you know? At that time, every male elephant becomes very arrogant. But Pratap, he was very loyal, very calm. My father said: I will call him at the time of breeding with a female elephant and see if he comes over or not. My father called him. He came over." Enamul, sipping on his Coke, took the moment to loudly belch.

The lords spoke to each other in Goalpariya for a moment, and I heard Enamul mention the BSF; he was telling the raja that I had been unable to get permission to visit the fence. The raja turned to me. "Do you want to see where Bangladeshis are living?" he asked. "It is very near." We marched down the stairs and down the hill and into the car, the raja riding shotgun. A suspect cluster of eight huts soon appeared at our right, on a patch of land that once belonged to the raja. "Slow. Slow. Slow." Enamul whispered. "Slow. Slow!" We craned our necks, but all there was to see was a woman in a sari who disappeared into one of the huts, carrying something. "You can't say for certain that they are coming from Bangladesh," Enamul said. A Bengali was a Bengali. "But you can't say for certain that they are Indian."

The land was low-lying, a dozen feet below the elevation of the road, and was interspersed with rice paddies. When it rained, it would be the first to flood. To my surprise, it was flanked by two BSF camps. "Actually, this is government land," said Enamul. "When someone occupies it, the government does not inquire. Citizens do not inquire. You cannot. Bengali, Bangladeshi—the language is the same. It is difficult, because the physical and biological patterns of Bangladeshis, the characteristics . . ."

"Cannot be detected," said the raja. "Cannot be detected!"

"You can't detect them," Enamul agreed.

DHAKA, NOT DHUBRI, was as far as most of Bangladesh's migrants could go. As in Senegal, the poorest did not have the resources to go any farther than the capital. Dhaka's metropolitan area had an estimated thirteen million people, and it swelled with half a million more *char* dwellers, cyclone refugees, and other newcomers each year—the highest growth rate on the planet. By 2025, it will be bigger than Mexico City or Beijing. New arrivals camped for days or weeks in rail yards and bus stations before moving into sprawling slums that did not appear on official maps. Men often found work as $2-a-day rickshaw drivers; the megacity was now thought to have as many as 800,000 rickshaws. The lucky ones got jobs in illegal, collapse-prone factories producing clothing for the rest of the world. In Old Dhaka after Atique and I returned from the south, I witnessed a traffic jam near a statue of a Kalashnikov rifle: rickshaw after rickshaw in a dead standstill that stretched for half a mile down a narrow street. The Dhaka sky was either smoggy or raining, and the city itself was putrid and strangely beautiful. "The capital of Bangladesh seems at times to be dissolving into its constituent elements," wrote the journalist George Black, who visited a year before I did. "If it's made of iron, it's rusting; if it's vegetable, it's rotting; if it's brick, it's reverting to mud, to river sediment."

No one I met wanted to discuss India's fence. "Please, we cannot talk about such things," said a representative of the International Organization for Migration (IOM). "It is very sensitive." The official Bangladeshi government position was that illegal migration into India did not exist. Instead, the IOM told me about legal migration: the Bangladeshis who paid upwards of $2,000 to recruiters to join the bottom rung of the labor pool everywhere from Malaysia to Dubai to Iraq. There were seven hundred recruiting agencies in Dhaka and as many as two million Bangladeshi guest workers around the world. They were found in the Maldives, one of the few countries on the planet as imperiled as Bangladesh by sea-level rise, and in Libya, where, during the revolution, they fled to Europe in small boats, along with escaping Africans. At least one fishing boat packed with hundreds of Bangladeshis landed in Malta.

Most aid workers in Dhaka did not want to talk about the fence, simply because it represented defeat. As a rule, the young Bangladeshis I met, unlike young Senegalese, did not dream of moving permanently to another country, however threatened their own might be. The national ethic, learned on the *chars,* was to keep on adapting. Atique's organization, COAST, preached more rice cultivation, more education, more local control. "Because of the food crisis, foreign companies are coming here and telling us that genetically modified hybrids are the only way," its founder told me. Years earlier, Monsanto had partnered with BRAC, the largest NGO in Bangladesh and now the world. "There is not an example anywhere of companies just doing good for the people, but that is not to say that we are against all foreign aid," he continued. "One Bangladeshi produces 0.3 tons of carbon a year. One American, 20 tons. We deserve the money."

Meeting with NGOs usually meant shuttling from my hotel near Old Dhaka to the comparatively tranquil enclaves of Gulshan and Banani, which had expatriates and trees and which required a taxi ride rather than a rickshaw. The six-mile trip lasted up to ninety minutes; Dhaka has the worst traffic I have seen anywhere in the world. Drivers let their vehicles grind up against one another as they jockeyed for position, and nearly every bus and every truck and two-thirds of the cars had long scrapes down both sides. Even in the spacious Gulshan offices of Atiq Rahman, the venerable environmentalist and IPCC author, the traffic noise didn't go away. "Migration is not adaptation," he told me the afternoon I visited, and then his voice was nearly drowned out by a chorus of horns. "For us, climate adaptation is modifying your systems"—honk, honk, honk—"through technologies, through assistance. The moment you have retreated, that is not adaptation." Honk, honk. "Migration is defeat. With climate change, there are three possibilities: Adaptation. Mitigation. And defeat."

Rahman was one of the few people in Dhaka willing to talk about the fence. Just as cattle rustling was inevitable—"We need cows," he joked, "and India has a lot of cows they don't need"—some migration was inevitable no matter how much Bangladesh could adapt. His only hope was that the defeat be well managed. He told me about a reception he had recently attended in Los Angeles. "I told the Americans: I want a piece of California. I want a piece of Texas. I want a piece of Maryland for my people that you are inundating," he said. "I can do the calculations, look into your emissions, to see how many each should take. I can determine how many Germany should take." What Bangladesh had was workers. What America increasingly had was old people. "Many of them will want to play golf," he said. "Many of them will have money, and they will need nursing help. They should have died at sixty-five, but they hang on until seventy-nine, and they'll continue to hang on until eighty-five. They will need their massages. Rather than do a threatening migration, I would try to turn the climate-change migrant into an effective, resourceful economic migrant—something for both sides." The alternative was bleak. "If that doesn't happen," he continued, "then I think we will run. And stop us if you dare. With what? How many bullets have you got?"

Despite the specter of climate change, Bangladesh received less foreign aid now than it did at the end of the 1990s—about $1.5 billion a year, a quarter of what its exported laborers sent back in remittances. The question Bangladeshis asked was not whether India would complete its fence—it surely would—but whether the high emitters of the West would make good on pledges of climate aid. Multimillion-dollar adaptation funds, filled largely with IOUs, were springing up on paper the longer international climate negotiations dragged on, yet some of the promised aid was simply existing aid, rebranded. Rahman could name only one project that had received funding: $200,000 for coastal reforestation work that would require $23 million to complete.

"There is no money!" he declared. "Money doesn't get to the poor. That is the nature of money." He explained how the Clean Development Mechanism, created by the Kyoto Protocol in 1997, allowed polluters in the developing world to get paid for reducing their emissions, earning China and India hundreds of millions of dollars. Bangladesh, with far fewer emissions to cut, had earned very little. It was a corrupt system, he said. The polluter didn't pay; the polluter got paid. Carbon got dispersed equally around the world, but reparations money, however well-intentioned, did not. Rahman became quieter, the traffic noise louder. "The nightmare scenario on climate change," he said, "is that there will be money floating everywhere. Floating around. A lot of money floating around, and a lot of zero-carbon technology being transferred to places that already produce virtually zero carbon. And nothing happens. For the poor, absolutely nothing happens."

"THE FENCE WILL not be enough to stop them," admitted Enamul before I returned to Guwahati, as we walked past cows on the riverbank. "But still we must complete it. Because otherwise there is nothing."

—————

I STILL WANTED to see the fence with my own eyes. In Assam's neighboring state of Meghalaya, I finally did. From Shillong, the state capital, once a British hill station owing to its five-thousand-foot elevation and temperate air, I drove across the plateau to Cherrapunji, which is listed in *Guinness World Records* for the most rainfall in a year and in a month (1861 and July 1861, respectively). Sign after sign announced, "Welcome to the Wettest Place on Earth." Along the sides of the road were chow mein shops owned by local Khasis, a Christian tribe that was fond of country music, especially love songs. I persuaded my taxi driver to follow the increasingly muddy road off the edge of the plateau, and down we dropped toward Bangladesh, losing forty-five hundred feet in less than an hour, skidding around corners, passing thin, horsetail waterfalls that tumbled down towering cliffs. The air became hot again, the people darker. Water trickled down to Bangladesh. People trickled up. This was how the world was.

The border, when we reached it at the end of the road, was anticlimactic: no Bangladeshis running across, no easy metaphors, almost nothing at all. There were a few wandering cows, a few homes and villages just yards from the zero line, a bucolic scenery of fields and farmers, and a pair of soldiers from Kolkata in a thatch hut. Their blank expression was the same I'd seen on the Vandoos on Devon Island, and their job was the same. They pointed their guns and waited. The fence was two rows of barbed wire separated by a narrow patch of no-man's-land, and it looked formidable, even impenetrable. You could almost believe that all problems would remain safely stuck on the other side.

TEN

SEAWALLS FOR SALE

WHY THE NETHERLANDS LOVES
SEA-LEVEL RISE

O n a Monday the year before Hurricane Sandy hit New York City, attorneys and ambassadors sat in a cavernous auditorium at Columbia University discussing what happens, legally speaking, when an island nation disappears under the sea. There were "novel questions," said the law professor Michael Gerrard in an opening statement. "If a country is underwater, is it still a state? Does it still have a seat at the United Nations? What becomes of its exclusive economic zone? Its fishing rights? Its rights to undersea minerals? Can its statehood be prolonged? What is the citizenship of its displaced people? What are their rights in the places where they will go—and who will have to take them in? And do the country and its people have legal remedies?"

The auditorium was shaped like a shell, and it had more vertical relief—perhaps fifty feet from lectern to cheap seats—than did many of the islands in question. More than two hundred people, most of them in suits, were packed in its ten rows. They included at least one lawyer looking for new business, or so she told me, and numerous representatives of AOSIS: the Alliance of Small Island States, a forty-four-nation bloc uniting global-warming poster children such as Tuvalu and the Maldives and seldom-discussed victims such as Grenada, Cape Verde, and the Bahamas.

Gerrard had convened the conference along with the UN ambassador from the Marshall Islands, one of the lesser-known AOSIS states, which consists of twenty-nine Micronesian atolls and five islands near the international date line in the middle of the Pacific Ocean. Though a bit player in the UN, the Marshall Islands have the peculiar moral authority that comes with twice facing annihilation from someone else's pollution: In the 1940s and 1950s, the country was better known as the Pacific Proving Grounds, where sixty-seven nuclear devices were detonated by the U.S. military. The world's first hydrogen bomb, Ivy Mike, was successfully tested here in 1952. ("It's a boy!" wrote the bomb designer Edward Teller in a telegram to Los Alamos.) Two years later, America's largest-ever blast, fifteen-megaton Castle Bravo, lit up Bikini Atoll. Now, thanks to their average elevation of seven feet and a high point of just over thirty feet, the Marshall Islands are expected to be one of the first nations extinguished by climate change. One islet is gone already: Tiny, verdant Elugelab was vaporized by Ivy Mike. Only a mile-wide crater remains.

The islands have had their independence from the United States only since 1986, their national anthem, "Forever Marshall Islands," since 1991. "With the light of Maker from far above," the anthem begins, its lyrics loaded with unintentional meaning. "Shining with the brilliance of rays of life/Our Father's wondrous creation/Bequeathed to us, our Motherland/I'll never leave my dear home sweet home." The local economy revolves around foreign aid, coconut farming, tuna processing, licensing of fishing rights, and service jobs at a remaining American missile base. The nation also sells hundreds of flags of convenience to ships trying to avoid regulations in their home countries; Shell's main Arctic drill ship, the *Kulluk,* bears the name of the Marshallese capital, Majuro, on its hull. The population of the Marshall Islands is sixty-seven thousand people, ten thousand more than that of Mininnguaq Kleist's wondrous Greenland—making it nearly a wash, from a utilitarian perspective, if the former sinks into the sea thanks to climate change while the latter gets its own independence.

The Marshall Islands and Columbia had begun planning the conference after the world failed to ink a new climate treaty in Copenhagen, after seeing the UN climate process, with its focus on multilateralism and emissions cuts, achieve so very little. "This conference is a recognition of a big failure," the ambassador said when it was his turn at the lectern. "There is no political will. No process. No urgency. A few weeks ago, in Bangkok, we spent one week discussing the agenda—one week discussing what to discuss! There is no light at the end of the tunnel, and that is why I approached Professor Gerrard." More was at stake than mere survival. "For us, our land and our natural resources—particularly marine resources—are part of the Marshallese collective identity," he said. Even so, retreat to higher ground was not an option in his country, he continued, and it would become uninhabitable long before full submersion: Occasional overwashing would dump seawater on cropland while contaminating the drinking water supply. While construction had begun on a seawall providing limited protection for Majuro, it cost a crippling $10 million per meter. Already dozens of low-lying homes had been flooded. Soon, the islands would experience a major outbreak of dengue fever—a mosquito-borne disease some scientists believe is worsened by higher temperatures and rainfall.

The first scholar to present research, tall and blond, stiffly recited the legal thresholds for statehood in a German-accented monotone. The first was a defined territory. This could still be met after sea-level rise by, for instance, artificial islands: floating structures towed into place, moored to the seabed. "Their capacity to generate [new] maritime zones was abrogated in 1958," she pointed out. (That is, they were useless for expanded Law of the Sea claims. Otherwise, the Arctic would already be full of them.) But as a replacement for existing territory, artificial islands might be accepted by a sympathetic or guilty world. The second threshold was a permanent population—fulfilled in the future, perhaps, by "a population nucleus, a legal anchor, a caretaker population" on certain islands. This was like Sergeant Strong's plan for Hans Island, expanded.

The third threshold—a government—was easy to imagine: like that of Tibet in Dharamsala, India, a government in exile. The last threshold, independence, "is de facto granted by the international community," she said. To kick a country out of the UN required a two-thirds vote. "I doubt that most of the members would vote to extinguish a small island state." If the world wanted to recognize the Marshall Islands even after they were unrecognizable, they could find a way to exist forever, at least on paper.

Another speaker introduced the concept of the "nation ex situ": the disappeared state as trusteeship, virtually there to take in reparations payments from the inevitable climate-change lawsuits—some of them no doubt filed by attorneys in this room—and to distribute them to the islander diaspora. The president of the Marshall Islands, a balding man with a blue tie, leaned forward in his seat at the front of the room and clenched his hands, as if bracing for a blow. The country's foreign minister stood up. "The wholesale relocation of our nation is no more acceptable to us than it would be to the countries of the many UN ambassadors in this room," he declared. Cape Verde's ambassador seconded him. "A lot of people think that sacrificed lands will die without shouting," he boomed. "But I assure you, we are shouting!" The issue, specified the Maldivian ambassador, "is not UN membership. Sure enough, these islands can disappear, and along with them a few millions of people, and the world will go on. But is that what human civilization has come to? When countries or peoples who become an inconvenience, that we let them go, as with the Darwinian survival of the fittest?"

The conference moved on to other sticky questions, starting with how an island nation could retain its fishing and offshore mineral rights when, according to the Law of the Sea, maritime territory is based on where your coastline is. If a country no longer had any coastline because it was underwater, would all be lost? An Australian professor, all waving hands and furrowed brow, offered the most convincing solution: Island nations should amend their domestic laws to use only geographical coordinates when defining their coastlines, not physical landmarks. Claim in the legislation that the lines will be periodically updated by further legislation, she suggested—"but then don't ever update it." And hope the world goes along with it.

The Columbia disaster expert Klaus Jacob, whose research on New York City's vulnerability to storm surges would turn him into a minor media star after Hurricane Sandy, presented his study of how long, and at what cost, the Marshallese capital would be sustainable. For planning purposes, he explained, "risk" was measured in dollars per year—a function of the annual probability of hazards multiplied by the value of the assets, multiplied by those assets' degree of vulnerability. Majuro had more than thirty thousand people packed onto 3.7 square miles of land with an average elevation below seven feet. "Looking at it," he said, "it's the many small events, not the occasional big event," that will contribute the most to losses. On average, the city would have to pay out 10 to 100 percent of the total value of all its buildings and infrastructure every year if seas were a meter higher—a crushing rate that no poor country, no matter how flush from litigation, could afford. "I think the only possibility is to reduce the population on Majuro," he concluded, "and just leave a steward population. And then, in a thousand years— that's how long it'll take for the seas to go back down—your people can go back."

The conference's most hotly contended issue was whether there should be a new climate refugee law, or whether existing refugee law could be expanded, or whether nothing could be done at all. "It is difficult to pinpoint climate per se as the reason for migration," noted an International Organization for Migration official. We learned that many thousands of Marshallese—perhaps a tenth of the islands' current population—had immigrated to Springdale, Arkansas. Springdale was the headquarters of Tyson Foods, the fast-food supplier and the largest meat producer in the world. Marshallese fleeing their country had landed there for slaughterhouse jobs that even Mexicans were beginning to eschew, and amid false accusations of bringing in leprosy and spreading tuberculosis, they now provided a new vital service for America: cheap chicken. But Springdale was an aside. Mostly, the scholars argued over precedents. The islanders got fed up.

"You know, this question reminds me of a story," said a Marshallese minister. "I was involved in the cleanup of Enewetak Atoll, one of the atolls used in the nuclear testing." In the late 1970s, the Americans scraped radioactive soil and debris off the surfaces of the surrounding islands and enclosed ninety-five thousand cubic yards of it in a bomb hole known as Cactus Crater. "They wanted to put up warning signs to warn people not to come to the island," he said. "They asked, what would you like us to write on these signs? My answer was: 'Don't ask us. You dumped it there. You ask yourself, and we'll put the words in there for you.'

"Now you ask us about the loss of a country," he continued, "the loss of a people, the loss of their culture and their identity. You ask us, what do you want to do? Where do you want to swim? Well, I don't know. But you tell us when to swim."

For the three days, this was the pattern: indignation from islanders followed by clinical detachment from at least some of the scholars, followed by more indignation, with many of the endangered people in the room—Maldivians, Bahamians, Micronesians, Nauruans, Saint Lucians, Palauans, Kiribatians—wincing every time one of the lawyers stated the meeting's premise out loud. The academics tiptoed for the most part, very aware they were charting a rational path to the unthinkable, and sometimes the gaps were visible anyway. "Perhaps more should be said about why to maintain the state," said a Law of the Sea expert from the Netherlands one morning. "If you want to convince the international community to maintain these states notwithstanding all the problems, you will have to indicate why, and for what purpose, and what the objectives are." He wondered what was wrong with two alternatives to the conference's convoluted legal solutions: that endangered nations make land purchases abroad—an option already being publicly discussed by some governments—and what he deemed even "more realistic," a simple merger of various Pacific states, including some with higher ground, into a new country. Everyone could just squeeze in together. "That would ensure many of the objectives," he said smugly. There was a momentary silence—perhaps a room contemplating whether the Netherlands wouldn't mind being united with Germany if it was convenient for the rest of us—and another scholar stepped in to set him straight. "From an ethical perspective," she began, "I think the answer is obvious. They are sovereign nations, some after a long struggle, and they want to maintain this."

THAT SOMEONE FROM the famously low-lying Netherlands could be so practical about sea-level rise was no surprise. The wealthy country's unlikely response to too much water mirrored that of others long facing a shortage of it: Experts in drought, Spain, Australia, and Israel, were not always happy about climate change, but they had found reason to improve on desalination plant designs, and now they were happy to sell you one; experts in flooding, the Dutch were not especially worried about climate change—and would gladly sell you a seawall. The Netherlands' history of battling land subsidence and reclaiming acreage from the boggy Rhine-Meuse delta dates back to the Middle Ages: The iconic windmills were used to power the pumps. Its faith in techno-fixes is grounded in a landscape that has become almost entirely man-made. Two-thirds of its population lives and 70 percent of its GDP is generated below sea level. In 1997, the country completed the $7.5 billion Delta Works: dikes, dams, and tidal barriers that make up the world's greatest coastal defense network, an engineering marvel far more complicated than any wall of trees or border barricade.

As the rest of the planet began worrying about the sea, the Netherlands aggressively promoted its water management expertise abroad, from dredging and engineering firms to amphibious architects. It could tout one prominent international success story already: Manhattan, home to Columbia and its legal conference, looked the way it did thanks in part to land reclamation by early Dutch settlers in what was then New Amsterdam. So long as the Dutch were in the Netherlands, Old Amsterdam could be expected to keep looking the way it did, too; the country's best advertising was its continued existence. In a testament to its seawalls, it was at the bottom of Maplecroft's Climate Change Vulnerability Index, near northerly Iceland, Denmark, Finland, and Norway: 160th out of 170 countries surveyed.

I had traveled to the Netherlands before the legal conference to understand how different climate change was for this wealthy country than it was for Bangladesh or the Marshall Islands. In Amsterdam at an event called Aquaterra, billed as the first gathering of the world's threatened river delta cities—from New Orleans to Jakarta to Ho Chi Minh City to New York—I watched the opening speaker declare that we were here because of a "new vision." "It's about adapting, about business development," she said, "about challenges and chances, value creation, solidarity, and being entrepreneurs!" In Naaldwijk, the site of the ten-million-square-foot FloraHolland building, where "fresh cut" flowers landed daily from Kenya, India, and Colombia for the world's largest flower auction, I got a well-worn tour of a state-of-the-art floating greenhouse. In the ancient city of Delft, elevation 3.3 feet, I visited a lab where "smart soils" were being developed to fill cracks in dikes and save the Netherlands from levee failures like those that drowned New Orleans during Hurricane Katrina. "The idea is to use bacteria to create bonds," explained a professor at the public-private institute Deltares. "You can create sandstone in a week, when it takes a million years in nature. With 100 billion bacteria, you can catalyze anything." They fed the bacteria urea—"It's what they like to eat," the professor said—and were trying to scale up the process through genetic modification. Smart soils could also be sold abroad. I would later see a Swedish architect propose using the technology as an alternative to the Great Green Wall: Dump some bacteria on the Sahara and simply freeze the dunes in place, in the process creating "antidesertification architecture" that could house climate refugees.

An hour from the Netherlands' reinforced coastline, amid the rolling hills and farmland, was one spot where the country was admitting defeat in its war against water. Yet even this would illustrate how different climate change was for those who could afford to adapt. Overdiepse was a teardrop-shaped, two-square-mile island of reclaimed land, or polder, as it is known, in the middle of the Maas River. The Netherlands is sometimes described as the drain for the rest of Europe, and in the case of the Maas, or Meuse, the flow comes courtesy of Belgium and France. As Europe's climate changes, the river is expected to swell with ever heavier rains. The Netherlands' response to higher water had always been higher dikes, but the national government had run the numbers and decided it could no longer keep up. Through a program called Room for the River, forty sites, including Overdiepse Polder, would be sacrificed—left mostly undefended as floodplains—so that more developed, less expendable regions could be protected. Among Overdiepse's eighteen farming families would be the first refugees in the world whose exodus could be so clearly attributed to climate change, or at least to fear of climate change.

One morning, the local Room for the River coordinator and I drove over a bridge that connected Overdiepse to the mainland and parked at a house belonging to a dairy farmer. Thirty-two years earlier, the smiling, spiky-haired man had been the first baby born on the polder, and now he sat here with his own baby and his two-year-old daughter, who offered me a wad of Silly Putty. His farmhouse would be demolished, the farmer said, along with the barn that housed his cows. But he was happy: His was one of nine families that would be allowed to stay on the polder, on an elevated mound built with government help, and this would give him room to expand his dairy. The other nine families were getting multimillion-euro payouts. One used the money to buy better land in North Holland, another to buy a farm in the south. One neighbor took the government money and moved to Canada, where there was plenty of open space and where the weather was getting better every year. "He already has ninety cows!" the farmer said.

Downstream of Overdiepse, dredging at one of the Room for the River sites had also threatened the habitat of a brown, eel-like, bottom-feeding indigenous fish called the weather loach. To satisfy Dutch environmental laws, a crew of four biologists, two hydraulic cranes, and one tractor working ten hours a day, five days a week, for six weeks had just relocated 1,636 weather loaches. They accomplished the task with a typical Dutch flair for engineering: They dammed off irrigation ditches in two-hundred-foot sections, pumped out nearly all the water, then sent men in with waders and nets. As a final step, they dumped loads of ditch sludge on dry ground and sifted through it by hand.

I asked the Room for the River coordinator how much money had been allotted to make the eighteen Overdiepse families whole again, and the answer was almost $140 million—by chance the same amount, give or take a few million, that the Netherlands had by then pledged in climate aid for the developing world. The budget for the entire Room for the River project was almost $3 billion—more than has ever been disbursed by all international climate funds combined.

BACK BY THE SEA, six miles away from the global headquarters of Royal Dutch Shell, I met another morning with a young architect whose target market grew the more nations like the Marshall Islands began to sink. Koen Olthuis was thirty-nine at the time and already hailed as a visionary. CNN and the BBC liked to quote him, and he once finished 122nd in a vote for *Time* magazine's 100 most influential people—out of the top tier, unfortunately, but above Katie Couric, above Osama bin Laden, and above Mary J. Blige. "There is life as we are used to it," he told me, "and we think we have to keep it exactly the way it is right now. But if we can change how we react to Mother Nature, then climate change is just another effect—it's a chance. I think too many people still see it as a problem. Of course, there are some problems, but let's focus on how it can improve our lives." He was tall, with a mop of wavy hair, and he wore standard-issue architect garb: black undershirt, black V-neck sweater, dark jeans, leather boots. As he talked, he looked out the big window of his office, a redbrick building just a few feet above a canal.

"We are lucky to be here in the playing garden of solutions," he said. Once, "this was a country as an empty painting. We filled it with roads, houses, and bridges, and we've kept filling it—the painting is never done." The Netherlands was winning a multifront war against seawater, rivers, and rain. "The big reason that other countries are interested in our solutions," he continued, "is that many cities, many big cities— almost 90 percent of them—are next to the water. They are next to a river, to an ocean, to a delta system. We're talking about New York, Tokyo, Singapore, et cetera. They are all in the same boat."

For years, the Netherlands had been playing defense. It erected barriers and pumped water out of the polders. Olthuis's vision, in brief, was to play offense—to build a floating world on top of the water rather than trying to keep the water out. Together with his development company, Dutch Docklands, he designed not houseboats but islands and infrastructure: highways, apartment buildings, parks, airports, churches, and mosques. He dreamed of floating and hybrid cities as big as 100,000-person Delft. "We are part of the climate-change generation," he said excitedly. "Architects and creative people should be the ones designing this new world. Others can always look back, but we need new ideas. This is our drive and our duty—we have to do it! If we don't do it, who else will do it? We should make it happen!"

To really work, a floating foundation had to feel just like what we already knew: solid ground. Rigidity was key. The larger it was, the more easily this could be achieved. "It will be exactly as you see out there now," he said, pointing out the window. He pulled out a sheet of paper and began sketching on it with a black marker. "With a houseboat, my house is here. And then I have to park my car over here. I have to walk. My children cannot play outside. But on islands, they can play outside, and I can park my car on it, and there are trees on it—that's what I like!" With luck and good lawyers, the hybrid cities of the future would sit partly upon Olthuis's proprietary buoyant foundations: modular, interlocking units made of foam and concrete and protected by a suite of international patents.

His designs were eminently exportable. "You can find sinking islands worldwide," he told me. "There are many, many island nations with this problem." Tuvalu and Kiribati could not expect to save themselves by ringing their various atolls with seawalls—as in the Marshalls, the scale and cost of it were inconceivable. But artificial islands, whether or not the lawyers said they were enough for future UN membership, held promise.

Olthuis was planning his first visit to the Maldives in little more than a month. The lowest-lying country in the world, consisting of more than a thousand islands and twenty-six atolls spread along a six-hundred-mile-long archipelago in the Indian Ocean, was an alluring opportunity: Its leaders believed strongly in climate change—they held a cabinet meeting in scuba gear six feet underwater as a publicity stunt before Copenhagen—and among AOSIS countries it was relatively wealthy, a vacation hideaway for the rich and famous. And the tourism economy meant that it was already feeling the effects of climate change. Hotel chains, like oil companies, have a longer investment horizon than many industries, often twenty to thirty years. They weren't eager to sink money into beach resorts, Olthuis said, whose beaches were eroding away.

He had a patent-pending solution: a floating beach. He pulled up a drawing on his PC and showed me how it could attach to an existing island, expanding the life of a resort. In the stilted language of the patent application, the concrete-foam design would entail "a floating base on which beach material, such as sand, is applied to form the beach, with the particularity that the artificial beach has a flexible mat that is at least partially underwater." Its key technology, I was surprised to learn, was bacteria-produced artificial sandstone, a smart soil like that developed at nearby Deltares. The market was so much more than sinking islands, Olthuis said: "Dubai has hundreds of kilometers of coastline that are constantly being eroded." For a short moment, I could almost imagine the global elite bobbing along on a sandy bed of urine-fueled, genetically modified bacteria.

On his PC, Olthuis flipped to an earlier project Dutch Docklands had been hired to develop in Dubai. Known as the Floating Proverb, it was part of Sheikh Mohammed bin Rashid Al Maktoum's famous Palm Islands, man-made archipelagoes that were themselves constructed in part by Dutch dredging and land reclamation companies. The eighty-nine floating islands of the Floating Proverb were to have spelled out a poem written by the sheikh himself: "It takes a man of vision to write on water/Not everyone who rides a horse is a jockey/Great men rise to greater challenges." Since the financial crisis, this and much else in Dubai had ground to a halt. Aerial photographs revealed that the once prestigious development the World, an artificial archipelagic Earth, was already losing its shape, its component islands falling back into the sea. But for Olthuis, Dubai's boom and bust had worked out fine. He had been paid to develop his vision, and whether or not building resumed, he was confident there was plenty of other water out there.

Under the header "Green IP," the Dutch Docklands Web site would soon show images of floating gardens, floating solar panels, and even a floating, water-cooled mosque. The company began marketing what it called Affordable H$_2$Ousing: the architects' solution to the lawyers' quandary. There was an uncaptioned photo of the floating prison at Zaandam. There was a quotation from Olthuis's partner, Paul van de Camp: "We told the president of the Maldives, we can transform you from climate refugees to climate innovators." Dutch Docklands and the Maldives would soon ink agreements for everything from floating villas to a floating marina. Greenstar, a two-million-square-foot floating garden island with shops, restaurants, and a conference center that was originally designed for Dubai, would be recycled and rebranded as a Maldivian national icon. "The green-covered, star-shape building symbolizes Maldivians' innovative route to conquer climate change," read the ad copy. "This will become the Number 1 location for conferences about climate change, water management, and sustainability."

Before I left his office, Olthuis led me downstairs to a private screening room, where we sat back in plush leather chairs to watch the Dutch Docklands corporate movie. He turned on the projector, and a disembodied male voice, clipped and Euro-accented, came in over the speakers. "They say we use only 10 percent of our capacity for thought," it intoned as electronica began pumping in the background, "and we know we use only 30 percent of Earth's capacity for life, for living. Well, it's time for all that to change—and it's the Dutch who are doing it." The screen filled with a shot of a roiling blue ocean. "Many centuries of living with water, much of it below sea level, have taught us all we need to know about controlling our wet environment," the voice continued. Images of floating highways, mosques, neighborhoods, and apartment blocks flashed before us. "Even when confronted with endless stretches of open sea, we are their master . . . It's all researched, tested, ready for takeoff. So just think of all that idle water in your community and, together with Dutch Docklands, start putting it to work. Because where there is nothing, anything is possible."

"That last sentence, I really, really like," said Olthuis. "Because there's all this water."

SOUTH OF OLTHUIS'S OFFICE, the greatest port in Europe, Rotterdam, the point of entry for most of the Continent's oil, was being turned into a showcase of the Netherlands' climate readiness. One morning I joined a group of a dozen mostly American urban planners on a tour led by local officials and the Dutch multinational Arcadis, a $3.3 billion, twenty-two-thousand-employee engineering firm that derived its name from Arcadia: in ancient Greek mythology, the nicest place on Earth. The company logo was a fire salamander, an animal at home on land or in water. The senior Arcadis representative with us was Piet Dircke, the director of its international water program. "I'm one of the people who's leading all of the Dutch efforts to get a position in the U.S. based on climate-change adaptation," he told me. "I'm trying to connect the international ambitions of Rotterdam to cities like New York and New Orleans and San Francisco."

What was billed as the Rotterdam Climate Proof tour started on land at the edge of the North Sea, at the crown jewel of the Netherlands' Delta Works, the Maeslantkering, an enormous storm-surge barrier at the mouth of the port. The barrier consisted of two curved, floating gates that swung closed, then sank into place when one computer system—known as the BOS—predicted a storm surge of at least three meters and told another system—BES—to enact the closure sequence. It was among the largest moving structures on the planet. Each steel swing arm was twice as long as the Statue of Liberty is tall. The Maeslant barrier took six years and $500 million to build and install, and when it was finally in place, Queen Beatrix of the Netherlands came herself to inaugurate it. Since then, it has been used only once, in 2007. It was built to withstand all but a one-in-ten-thousand-year storm, though climate change, we were told, could muddy the math.

The planners on the tour busied themselves taking photographs of the open swing arms and clambering up what passed for a hill to try to get an angle on the whole Maeslantkering. It was impossible. It was too big. Inside a visitor center, a guide showed us around a scale model, and after we were suitably impressed, we left for the center of the city. Near the historic headquarters of the Holland America Line, we boarded a water taxi, and soon we were cruising through a four-thousand-acre expanse of former shipyards that had become one of the largest development sites in Europe.

"Rotterdam's ambition is to be one of the places where the new future will be created," the port's redevelopment manager said after we'd clambered out to join him on a pier. We followed him into an ornate building that once belonged to the RDM shipyard. "RDM" no longer stood for Rotterdamsche Droogdok Maatschappij, he assured us. It meant Research, Design, and Manufacturing, and the site was being retooled as a futuristic campus for research institutes and technical universities attempting to solve the problems of the world. In adjacent buildings, students were perfecting zero-emissions go-karts, conversion kits for hydrogen-powered buses, and the Sustainable Dance Club, where the power of dancers' footwork kept the lights on. "We'll be a center for water tech and for clean tech—the Silicon Valley of the lowlands," he continued. "People will come even when sea levels are rising." A planner from the San Francisco Bay Area raised his hand. "But why invest here rather than somewhere else, like Singapore, Shanghai, or the Silicon Valley?" he asked. The port manager smiled. "Because we are turning a threat into an opportunity," he said. "We're sending a message to the international community: If you set up shop here, we can guarantee that we will keep your feet dry."

Having decided to be the global leader in climate adaptation and water knowledge, Rotterdam had created a network it called Connecting Delta Cities, hosting conferences and hastening the flow of its experts and expertise to member cities on six continents. Across the harbor from us, firms including Shell, BP, IBM, and Arcadis had been persuaded to take part in the Rotterdam Climate Campus. "It will likely be floating," an official told us. Elsewhere in the harbor would be floating neighborhoods and floating laboratories, and some RDM students would be housed on the SS *Rotterdam,* the onetime flagship of the Holland America Line. "We even have a floating prison," someone said.

Some innovations were already springing up abroad of their own accord. In New Orleans, Brad Pitt's Make It Right foundation and the Los Angeles architecture firm Morphosis would unveil the Float House, which could rise up to twelve feet on guideposts as floodwaters destroyed its neighbors. Pushing a similar design was the professor behind the Buoyant Foundation Project, who described her areas of research as "the study of wind loads on tall buildings, the aerodynamics of wind-borne debris, strategies for the mitigation of hurricane damage to buildings, and the origins of early 20th-c. Russian avant-garde architectural theory in 19th-c. mystical-religious slavophile philosophy." But when it came to seawalls, storm-surge barriers, and other city-scale defenses, firms like Arcadis couldn't help but believe that their services were needed. "The consequence of 'New Orleans' is that the Americans have placed orders with a number of Dutch companies to the value of 200 million dollars," read a quotation from Piet Dircke in one of the Port of Rotterdam's pamphlets. Arcadis had seventy-one projects in the New Orleans region alone, including part of the two-hundred-foot-wide Seabrook Floodgate, a mini Maeslantkering. And Dircke, I learned, had been to New York City four times in the previous six months.

Dircke and I sat together on the water taxi ride back. "Of course we build on our Dutch reputation as that very small but very brave country battling for centuries against the sea," he said. "Climate change brings opportunities. You get new challenges." He brought up the Elfstedentocht, the Netherlands' famous speed-skating competition, lamenting that ice-skating was becoming an indoor sport. "Are we not living in a crazy world? I'll tell you what is even more crazy: You can do pretty good skiing now in Holland. It's down in the south, and it's called Landgraaf: an indoor ski area. Everybody was laughing about it until two years ago, when they opened the World Cup ski season there. There was no snow in the Alps, and Landgraaf had snow. You know what happened after the World Cup? The Austrian and Swiss teams quickly booked training periods for the next year. In Holland! Imagine the world, a couple years from now, when we have only indoor skiing and no snow on the mountains. And it seems we are already adapting to it. It is normal." He chuckled. "We are already adapting. Our minds are adapting."

Dutch companies had already helped build storm-surge barriers for Venice, New Orleans, London, and St. Petersburg—metropolises that could afford to pay much more than any island nation could—but increasingly they looked at New York City. The task would be complex and lucrative. "You can't seal off New York with just one barrier," Dircke said. "You need an East River gate. On the New Jersey side you need a gate. At the Verrazano Narrows you need a gate. And you need a gate near Jamaica Bay if you also want to protect JFK airport. There are four holes, luckily not more than that."

It was three years before Hurricane Sandy began forming in the southern Caribbean and spinning its way north. A conference had just been announced by the American Society of Civil Engineers (ASCE) to look at some of the first designs for a New York storm-surge barrier, and Dircke would soon be on his way to the city to present Arcadis's idea. "It is very exciting," he said. I decided to follow him there.

ANOTHER BOROUGH, another auditorium, another conference on sea-level rise. This time it was not at Columbia in upper Manhattan but in the decidedly less imposing environs of New York University's Polytechnic Institute in downtown Brooklyn, not far from where insurance companies were quietly dropping clients near the Gowanus Canal. The ASCE's "Against the Deluge" conference was the rare scene where the line for the men's bathroom was always far longer than that for the women's. It had the air of what was then a lost cause. There was a single paying exhibitor—"Please visit our exhibitor," implored the organizers—who was an eager Texan waving flyers at all the old men as they waited for the provided spaghetti dinner, which was served cold. The Texan's invention, FloodBreak, was an ingenious, self-deploying floodgate that was big enough to protect one's garage but unfortunately not at all big enough to protect Manhattan. Inside, scientists explained the growing threat to New York: The one commonality with Bangladesh was that sea levels here were rising faster than the global average, a foot in the last century. This rate could double just as the city faced more powerful hurricanes. At risk, a city employee told the half-empty room, were 802,000 buildings worth $825 billion with contents worth $560 billion. Another speaker highlighted Breezy Point, the Queens neighborhood that would be flattened by waves and fire in Hurricane Sandy, as particularly at risk. The scientists were followed by engineers and architecture firms presenting rival storm-surge barrier designs, and the Arcadis team gave a sales pitch that was more subtle than the Texan's—and much more effective.

Dircke's proposal for the Narrows was a beautiful design—"An extra landmark for New York," said a colleague—that combined the Maeslantkering with two other famous barriers in the Dutch Delta Works, the Hartel and the Eastern Scheldt. It would allow passage of the biggest ship in the world, the 1,300-foot-long, 185-foot-wide *Emma Maersk,* while also protecting the most moneyed place in the world, Wall Street, from a twenty-two-foot surge. Not including the other three barriers the city would need to be fully sealed off, the design would cost a very roughly estimated $6.5 billion, more than double the price of the Netherlands' Room for the River project. The Arcadis presentation featured spinning animations of the gate action and an aerial glamour shot of the New York Harbor of the future, safe behind its floodgates and beneath a sunny sky. When it was over, the assembled engineers gave a rare applause.

There was one downside to any design that used the Narrows to protect Manhattan from storm surges—a necessary evil—and Dircke was straightforward about it. As everyone knows, when water is blocked, it doesn't just disappear. It flows elsewhere. If a surge came barreling toward a Verrazano barrier, it would do the hydrological equivalent of a bounce, and it would land somewhere else. Arrochar and Midland Beach on Staten Island, Bath Beach and Gravesend in Brooklyn—these and other immigrant-heavy neighborhoods were just outside the Narrows, poorer than the core, just above sea level, and slated for an even bigger surge. Manhattan would be saved, and they would likely be underwater.

When Hurricane Sandy hit New York City in late October 2012, there was not yet a barrier, just a hint of what could come. On Staten Island, a sixteen-foot storm surge swamped Midland Beach and Ocean Breeze and Oakwood Beach, and twenty-three people died, more than in any other borough—the vast majority of them south of the Narrows, the vast majority of them by drowning. In lower Manhattan, water flooded subway tunnels and power stations, and the cityscape went dark, with one exception: At 200 West Street, close to the island's southernmost tip, Goldman Sachs headquarters was ringed with a massive wall of sandbags, and backup generators kept the lights on all night. Across the stormy Atlantic, in the Netherlands, Arcadis's stock jumped 5.6 percent, capping a 43 percent rise for the year.

ELEVEN

BETTER THINGS FOR BETTER LIVING

CLIMATE GENETICS

The yellow-fever mosquito, *Aedes aegypti*, better known today as the primary carrier of dengue fever, is a container breeder. It lays its eggs in the pools of rainwater left in the things we leave outside our homes: buckets, vases, cups, yard ornaments, clogged gutters. The most proven way to eradicate the disease is to clean up those things or to constantly dump out their water, which for public health authorities is a grueling, house-to-house, yard-to-yard fight. And the more plastic human life becomes, the more our detritus becomes mosquitoes' habitat, the harder dengue is to control. There is still no vaccine for it, and *Aedes aegypti* bite mostly by day, making bed nets largely useless. Urbanization, globalized trade, and increasing air travel have also helped grow dengue into a global epidemic three thousand times more prevalent than it was in the 1960s: every year, as many as a hundred million infections and twenty-two thousand deaths in more than a hundred countries. *Aedes* like it hot, and they prefer humans to any other animal. They are attracted by the CO_2 we exhale with every breath, and their potential range expands, many scientists believe, with every ton of CO_2 our industries emit.

Aedes aegypti is originally from Africa, and *Aedes albopictus,* the other species that can transmit dengue, is from Asia. One or both are now found in twenty-eight U.S. states, Florida chief among them: In 2009, the first American dengue outbreak in seventy-five years took place in Jimmy Buffett's and Ernest Hemingway's paradisal Key West, the southernmost and hottest city in the Lower 48. A tourist had returned sick to New York, and soon the disease was tracked to a quiet street in Old Town. There were twenty-seven confirmed cases that year, sixty-six the next. A CDC team took random blood samples and estimated that 5 percent of Key West's population—more than a thousand people, many of them asymptomatic—had been exposed to dengue. In its mild form, the disease causes headaches, fevers, rashes, bleeding gums, and intense joint and muscle pain. Its more severe form, known as dengue hemorrhagic fever, brings nosebleeds, purple splotches under the skin, and possibly death.

Pending regulatory approval, Key West would soon also be the site of the United States' first release of genetically modified (GM) mosquitoes. The flagship product of a British company called Oxitec, the patent-protected *Aedes aegypti* OX513A was a kind of Trojan horse. Sent out by the millions to breed with native *Aedes,* the modded mosquitoes carried a suicide gene that would theoretically doom the next generation to an early death, making dengue transmission impossible. Genetic modification was the logic of climate adaptation taken one step further: Instead of changing how and where life was lived, it would change, however modestly, what life was.

I visited Key West in August, when the weather is at its stickiest. With the air-conditioning on max, the Florida Keys Mosquito Control District inspector John Snell drove me one morning to the highest point in Old Town, elevation eighteen feet above sea level, and parked his pickup. Nearby was the city's historic cemetery: nineteen palm-dotted acres where tourists came to see the grave of the forty-inch-tall midget "General" Abe Sawyer, who was buried in a full-size plot; the headstone of a hypochondriac waitress named B. Pearl Roberts, which read, "I told you I was sick"; and the resting place of the nurse Ellen Mallory, who treated yellow fever victims in the early nineteenth century, decades before anyone connected the disease to mosquitoes. While not the island's original cemetery—that one was destroyed by a colossal 1846 hurricane, its bodies left strewn in trees—it was old enough to be the most difficult part of Snell's beat. The graves of nearly 100,000 people, four times the island's living population, translated to lots of fresh flowers. "Some of the vases are just a constant, constant battle," Snell said. "The ones I can tell have been there a long time, I'll just go ahead and dump them out. But fresh ones I treat, half a larvicide tablet in every vase." He went through two hundred tablets a month.

Snell was one of eight home inspectors in Key West—twice as many as the city had before the dengue outbreak. He wore wraparound sunglasses and a white collared shirt, and when we left the truck, he was carrying a jury-rigged ski pole–water scooper in his hand, larvicide and a turkey baster in his black fanny pack. He was tasked with ridding some forty blocks and eleven hundred homes of *Aedes,* and his workload ebbed and flowed with the seasons. Warmer temperatures accelerate not only mosquito development but the incubation period of the dengue virus; inspectors have a shorter window to stamp out both host and disease. "In winter, in the dry season, it's not that bad," Snell said. There were two weeks to catch mosquito larvae before it was too late. But in the hot, humid summer, he had only four days.

"This is the big problem right now," he said as we approached a dilapidated fence. "There are a lot of foreclosures. And once a house goes into foreclosure, the bank shuts down the pool service and the landscaping and whatever else they have going on, and things just go bad." A trashed-out yard was ideal for breeding *Aedes,* and Florida vied with other Sunbelt states—Nevada, Arizona, California, Georgia—for the highest foreclosure rate nationwide. Even wealthy Key West had its share. Meanwhile, homeowners' insurance rates, higher than in any state but Katrina-walloped Louisiana, were only rising as insurers pulled back from the coastline or exited the state entirely. You could not buy a home without additional windstorm and FEMA flood insurance, which often cost more than the main policy. You could not buy windstorm insurance except through the much-reviled, state-backed Citizens Property Insurance Corporation, the onetime insurer of last resort that was now Florida's largest as it absorbed policies from private firms that had fled the state. The Caribbean is expanding only slightly less rapidly than other seas, and Key West—which has the longest sea-level record in the Western Hemisphere—was overdue for another 1846-style hurricane.

Snell placed his hands atop a locked gate and jumped over in one easy motion. On the other side were a wooden deck, a palm tree, a small swimming pool, and a Jacuzzi. The heat was suddenly stifling. "There could be a twenty-knot wind, and it's completely still in these yards," he said. He had appropriated the Jacuzzi to raise fish for his fight against dengue: small, larvae-eating gambusia, which he would release in water cisterns and bird feeders as he traipsed through the backyards of paradise. Keys residents had historically kept cisterns—ideal *Aedes* breeding sites—under their homes, and more than 350 of them remained, along with nearly 250 wells. Scientists had determined that if less than 2 percent of homes contained *Aedes,* that was good enough: There could be no dengue transmission. But that summer, two Key West neighborhoods had indexes approaching 50 percent. Snell looked around the yard for signs of mosquitoes, and finding none, we jumped back over the fence. In the street, lined with pastel homes with their shades drawn, their owners gone for the summer, we saw not a soul.

It was extremely difficult to model dengue's spread, Snell's boss at the Mosquito Control District, Michael Doyle, told me. There were too many factors. This was especially true when gauging the effects of climate change. Big storms could lead to breeding sites in the rain-soaked debris—witness the *Aedes aegypti* explosion in the Cayman Islands after 2004's Hurricane Ivan—but droughts could be equally dangerous if people began storing extra water in open containers. "It's not the simple connection that if it gets warmer, this mosquito will be everywhere, that it will just move north," he said. "It's also how weather affects humans, you know? If it's really hot, people may spend more time inside, where there's more air-conditioning going on, so there's less contact with mosquitoes." Doyle and his extended family had just moved here from Colorado, where he had battled West Nile virus, another mosquito-borne disease linked to climate change. Already his mother-in-law was complaining about the mosquitoes at their Keys rental house, so his new employees were preparing a special strike team to stamp out the problem.

Until federal regulators decided the dengue threat was severe enough to let in Oxitec's *Aedes aegypti* OX513A, the Mosquito Control District would have to back up inspectors like Snell with another kind of air support: a sprayer-equipped Bell 206 helicopter that hovered fifty to sixty feet above Old Town twice a month, raining insecticide onto rooftops and tourists' rental cars. The insecticide, VectoBac, was based on a strain of the natural bacteria *Bacillus thuringiensis* (Bt) and, explained Doyle, killed mosquito larvae but little else. On a newly waxed car, its droplets looked like dried milk.

I had timed my visit to watch the helicopter spray. Doyle and I met at dawn the next day to follow twin contrails of insecticide through Old Town. The helicopter had to cover an area of 950 acres, and its hundred-gallon tank could do just 200 before a refill. The pilot rushed through five sorties as quickly as possible, lest his work be ruined if the wind picked up or the humidity dropped, lest the district be billed more helicopter time than it could already barely afford. Our SUV drove slowly through the backstreets, catching glimpses of contrails partly obscured by telephone poles, roofs, and wires. Only when we steered to an open patch of scrub alongside a busy road, across from a Lutheran church, did we have a clear view of it racing to and fro. We stepped out into the sun, and Doyle began telling war stories, like the one about the suppression campaign in Colorado when they'd carried their insecticide in backpacks and hand sprayed an entire forest. "Thirteen guys with thirteen backpacks," he said. "All scratched up, all dirty. And we did 56 acres!" The helicopter made a beautiful, wide-arced turn above the church and barreled back toward us. We retreated into the SUV. On the street, a homeless man walked by pushing a bicycle. He peered into the sky, covered his mouth and nose with an old T-shirt, and kept walking.

Dengue had hit Key West just as the conservative Florida legislature had limited local governments' taxing authority. The Mosquito Control District would spend almost $12 million in the 2011–2012 fiscal year but take in less than $10 million. It was burning through its cash reserves. And as aerial combat went, helicopters were much more expensive than transgenic mosquitoes—another reason the district really wanted them to be approved. Oxitec really wanted approval, too: It had already paid $130,000 in lobbying fees to the Washington, D.C.–based McKenna Long & Aldridge—Monsanto's sometime law firm—but had yet to see results. How the public would react seemed scarcely considered. When I arrived in Florida, outreach about the impending cloud of OX513As had so far consisted of a single presentation to the local gay business alliance. (Soon, after attacks by Friends of the Earth and other anti-GM campaigners, the news was splashed across the front page of the Mosquito Control District Web site: "Special Notice. Genetically Modified Male Release Trials.") In her presentation, the district's representative had explained that hundreds of thousands of "sterile" Oxitec mosquitoes would be released every week for six months. Only female mosquitoes bite; these would be males. A cocktail of inspections, insecticides, and OX513As would reduce the native *Aedes aegypti* population "to zero or near zero." Sustained, low-level releases would keep it down thereafter for a fee of $200,000 to $400,000 a year. Of course district staff would remain out in force, but to truly beat dengue, they would have to harness nature itself, at least a kind of nature. As she awkwardly explained to the gay business owners, male mosquitoes "are more effective than humans at finding females."

"THE KEY ISSUE," said Oxitec's founder, Luke Alphey, "is that we need to get enough wild females to mate with them. It's a question of quantity and quality. This is the quality test: Are the males sexy? Are they fit? Are they healthy? Are they happy?" In mosquitoes, he said, there were indirect proxies for determining this. Longevity was easily measured, and unfit *Aedes aegypti* died young. Size mattered: Small mosquitoes have smaller energy reserves. Maybe there was something to symmetry. Attractive humans were symmetrical. "Kylie Minogue has a symmetrical face," he told me when I visited him in England. But the only way to really know if females would accept a genetically modified male instead of the natural variety was to do field tests, and that was why dengue's expansion to GM-friendly places like Key West was so important to Oxitec.

Neither Alphey nor his company oversold the disease's complex link to climate change, but Oxitec's Web site sent visitors to a report by the Natural Resources Defense Council singling out global warming as a major factor in dengue's global expansion, and its own pages also highlighted warming's impact. "With the progression of climate change and the globalisation of travel and trade," read a section marked "Epidemiology," "it is predicted that dengue fever may spread further outside the current tropical zones." Climate change was, at the very least, yet another reason that the world might want to buy Oxitec's products.

Alphey's office was on the second floor of a brick building covered in wild grapevines, surrounded by a well-kept lawn and a copse of trees at the edge of an industrial park a dozen miles from Oxford University. The office itself was modest, mostly undecorated but for a scattering of papers, and Alphey, forty-seven, was tall and fit looking. He had a moderately symmetrical face. While activists have attacked Oxitec for its perceived secrecy, with me the former Oxford professor was an eager teacher, happy to spend the morning explaining the science behind his crowning invention.

He called it RIDL: "release of insects carrying a dominant lethal." Protected by U.S. Patent application 11,733,737 ("the invention relates to a non-human multicellular organism carrying a dominant lethal genetic system"), it was, in Alphey's explanation, a new way to carry out what was an old method of bug suppression. In the 1950s, entomologists had pioneered the Sterile Insect Technique (SIT): irradiating lab-raised fruit flies or tsetse flies, then releasing them. They mated with wild females but could produce no offspring. Unfortunately, mosquitoes were too fragile for SIT; the radiation killed them. So Alphey had sought a way to bake auto-elimination into their genes. He found it in a synthetic DNA known as tTA—a fusion of DNA segments from the bacteria *E. coli* and the herpes simplex virus—which he soon began inserting into *Aedes aegypti*. One difference between Alphey's technique and traditional SIT was that the mosquitoes it created were not technically sterile. They could mate and produce offspring, but these would not grow past the larval stage without the presence of an antidote, the common antibiotic tetracycline. In Oxitec's mosquito nursery, there was ample tetracycline. In nature, in theory, there was not.

In one study to test RIDL, Alphey had placed OX513As in one set of cages, unmodified males in another, and thrown in some "wild type" females. The OX513As were clumsy: They inseminated just over half as many partners, possibly because they ran out of sperm, and, unlike their rivals, seemed unable to distinguish between virgin and sullied wild-type females. But over a short time frame, three days, the modified and unmodified lines performed equally well. For investors, this might have looked like a silver lining: Not only would Oxitec have to produce and release swarms big enough to compete with the native population; it would have to do so quite often. A good rule of thumb, Alphey said, was twenty modded mosquitoes per week per human. "For a city of 5 million people," he wrote in a paper, perhaps imagining Miami or Madrid or Ahmedabad or Belo Horizonte or any number of second cities in the third world, "this would correspond to releasing 100 million males per week."

To those wary of genetic modification, intentionally releasing a transgenic organism into the wild might seem far scarier than optimizing an already domesticated crop. This is what the agricultural behemoth Monsanto, the world's biggest seed company and first name in gene engineering, does over the howls of activists. But products like Monsanto's—supercotton, supercorn—were designed to outcompete traditional varieties, Alphey pointed out. They were built to live. Oxitec's products, on the other hand, were built to die. "Self-limiting is much better politically," he told me. "You can say to regulators: If I stop releasing it, it all goes away."

Yet Oxitec's first *Aedes aegypti* field test, in the Cayman Islands 360 miles south of Florida, had been hugely controversial. The precursor to trials in Malaysia and Brazil and planned trials in Panama, India, Singapore, Thailand, and Vietnam, in addition to Key West, it began with local scientists manually separating male and female larvae by size—the females are bigger—using what one called "a sieve-like method." They achieved 99.55 percent accuracy, and three million OX513As were released in a forty-acre area. Another way to say that is that one-half of 1 percent of those released, nearly fifteen thousand mosquitoes, were genetically modified females capable of biting local islanders who had little knowledge about the experiment. But the results, published in late 2011, were impressive: After six months, the number of wild *Aedes aegypti* was reduced by 80 percent—"a complete success," proclaimed Alphey at a meeting of the American Society of Tropical Medicine and Hygiene, where he first announced the tests to a surprised world. (A later test, in Brazil's Bahia state, would reduce the wild population by 96 percent.)

In the limited public outreach by Cayman Islands authorities before the trials—leaflets, a five-minute promotional spot on local television—there was no mention of genetic modification. The mosquitoes were repeatedly described as "sterile males," language Alphey himself used until criticized for it. "If a female mates with a sterile male," read a 2010 joint press release from Oxitec and Caymans scientists, "she will have no offspring, thus reducing the next generation's population." Researchers at the U.S. Department of Agriculture and Germany's Max Planck Institute soon studied Oxitec's papers and regulatory filings and pointed out an issue that was more than semantic: In the lab, nearly 3.5 percent of the larvae born to a modded male and wild female somehow survived, even without tetracycline. Nearly 3.5 percent of 100 million mosquitoes is a big number. "There is the plausible concern," they wrote, "that females could inject tTA—the fusion of *E. coli* and herpes DNA—"into humans."

Alphey readily conceded one worry expressed by critics: If *Aedes aegypti* is wiped out, might not *Aedes albopictus,* the Asian tiger mosquito, come fill its ecological niche? "In places where you've got both," he said, "you've got to assume that by eliminating this one, you expand the other one a little bit. But *albopictus* is just a much less effective dengue vector." In some cases, he suggested, Oxitec's campaign against *aegypti* could happily morph into a campaign against *albopictus*—a kind of entomological Forever War. Oxitec's first RIDL prototype, OX3688, had in fact been a strain of *albopictus* developed as that mosquito expanded across the U.S. market. It was now in the "product optimisation" phase.

ONE PROMINENT SUPPORTER of GM mosquitoes is the $33.5 billion Gates Foundation, the world's largest charitable organization, which shares its founder's focus on techno-fixes. In his 2012 annual letter, Bill Gates declared that "innovation is the key" but noted a structural problem: If profit is the motive, third world problems rarely receive first world solutions. "The private market does a great job of innovating in many areas," he wrote, "particularly for people who have money. The focus of Melinda's and my foundation is to encourage innovation in the areas where there is less profit opportunity but where the impact for those in need is very high."

The Gates Foundation is so big that it can seem to single-handedly dictate global aid priorities, and two of its favorite causes are mosquito-borne illnesses and agriculture. In 2005, it gave a $19.7 million grant to a mosquito-modding consortium that included Oxitec and a number of public universities. (The money went toward open-source mosquito varieties, not OX513A.) The foundation also gave $13 million to a group in Asia and Australia trying to infect *Aedes* with a dengue-zapping bacteria, $62 million to the long-stalled international Dengue Vaccine Initiative, and half a billion dollars and counting to a partnership with GlaxoSmithKline (GSK) to hasten a long-awaited malaria vaccine that had been neglected—like many candidates for a dengue vaccine—because there had been little profit in a disease of the tropical poor. Like dengue, malaria was on a global expansion. "I think it's fair to say that all of us in the global health community are aware of the potential impact of global warming," says Dr. Rip Ballou of the Gates Foundation and formerly GSK and the vaccine's champion for thirty years, "especially so when it comes to diseases transmitted by vectors."

In agriculture, the equivalent of the GSK partnership was the Gates Foundation's collaboration with Monsanto—the emerging leader in the race for climate-ready crops and, like GSK, a publicly traded company that couldn't otherwise justify products meant for people who couldn't afford them. Monsanto has revenues of $11 billion a year, and its stock is held by everyone from Deutsche Bank's climate funds to the Gates Foundation itself. It was a subcontractor for the Gates-funded African Agricultural Technology Foundation, which had received $40 million to develop drought-tolerant corn for five sub-Saharan countries. In 2009, the continent's first varieties were tested under the South African sun.

Just after Christmas two years later, Monsanto's MON 87460—also a genetically modified, drought-resistant variety of maize—was quietly deregulated for use in Iowa, Indiana, and Nebraska. It came with a catch: The USDA had found it scarcely more drought tolerant than existing breeds. "Equally comparable varieties produced through conventional breeding techniques are readily available," read the environmental assessment.

Since 2008, anti-GM activists have tracked preparations for global warming by Monsanto and what they call the five other "Gene Giants": BASF, DuPont, Bayer, Dow, and Syngenta, the latter the alma mater of numerous senior Oxitec employees. The activists have identified at least 2,195 patent filings related to "abiotic stress tolerance"—resistance to extreme temperatures, resistance to droughts, resistance to anything in the environment that is not living but not friendly.

Dominating the climate-patent race were Monsanto and BASF, partners since 2007 in "the biggest joint biotech R&D program on record"—an eventually $2.5 billion effort to develop stress-tolerant corn, soybeans, wheat, cotton, and canola. Monsanto long ago shed its business as a manufacturer of chemicals—Phos-Chek for fires, Agent Orange for forests, DDT for insects—in the decades following a breakthrough: In 1982, its scientists were the first in the world to genetically modify a plant cell. But the patents upon which an empire was built—for the weed killer Roundup and for crops resistant to it—were beginning to expire. Monsanto needed another breakthrough. It was trying to reinvent itself. "How can we squeeze more food from a raindrop?" asked the Monsanto ad that appeared prominently in *The New Yorker,* *The Atlantic,* and *National Geographic.* When Monsanto and BASF identified a useful gene sequence in one plant, they often filed for a patent on it that applied to multiple plants. One issued to BASF in late 2009 is representative. "We claim . . . a transgenic plant cell transformed with an isolated polynucleotide," begins U.S. patent 7,619,137. The plant cell was found in any of the following: "maize, wheat, rye, oat, triticale, rice, barley, soybean, peanut, cotton, rapeseed, canola, *manihot,* pepper, sunflower, tagetes, potato, tobacco, eggplant, tomato, *Vicia* species, pea, alfalfa, coffee, cacao, tea, *Salix* species, oil palm, coconut, perennial grasses, and a forage crop plant."

"The more we know about the biology of the plant," Monsanto spokesperson Sara Duncan told me, "the more we pave the way for future advances." For biotech companies, the field of genomics—sequencing an organism's full DNA—provided a kind of real estate map. Rice was the first cereal crop and second plant overall to be sequenced, in 2005, five years after a rough draft of the human genome was assembled. It is relatively simple, a Rosetta stone for crop genomes, and lessons learned here can be applied to more lucrative corn and wheat. This is why three-quarters of it was already named in U.S. patent applications as of 2006. And why in the BASF-Monsanto collaboration that was becoming a climate-ready juggernaut, rice was the model crop.

Out of more than thirty-five hundred mosquito species, the second to have its genome decoded was *Aedes aegypti.* The first, in 2002, was *Anopheles gambiae,* one of sub-Saharan Africa's deadliest carriers of malaria and an important target of the Gates Foundation. After a researcher discovered that *Anopheles gambiae* is attracted to foul odors, the world's richest foundation once spent $775,000 to test traps that smelled like human feet and Limburger cheese. The Gates Foundation has notably spent not a penny on helping the world cut carbon emissions. "We believe the best way for the foundation to address climate change is to help poor farmers adapt," read an overview of its agricultural strategy. A stalk of GM rice seems nothing like a seawall, but to a technocrat it is the same—another patch, another software update for a world increasingly programmed by us.

OUTSIDE THE GREENHOUSE that was one of the BASF-Monsanto collaboration's principal pieces, it was Belgium in wintertime when I arrived, but inside the twenty-six-thousand-square-foot space it was equatorial and hot—Deutsche Bank's Wall Street tent all over again, only without the anaconda. The temperature was between eighty-two and eighty-six degrees, Marnix Peferoen told me as he removed his sweater, and the humidity was 70 percent. To my urban nose, it smelled rather like a brewery. The rice plants, in clear, plastic pots, each equipped with its own bar code and RFID transponder, were arrayed in perfect rows below thirty-thousand-lux lamps. The greenhouse was mostly empty of people, but a constant stream of Europop, synthetic and over-produced, blared from suspended speakers as we approached what was known as the Walking Plant System. Conveyer belts snaked through the building—"The same belts as in your car," Peferoen said—jerkily moving seedlings from one side to another as robots snatched some out of the passing parade. There were more than fifty thousand plants, and they would be here until they flowered—three to four months. They had probes to measure their water levels and tags to mark their age. The level of drought stress was seen in the plants themselves: dark green healthy, light green thirsty. "We mostly hit them with drought at the time they're flowering," Peferoen said, "but we can also decide to keep water low the whole time." In the summer, he said, the greenhouse's blinds were drawn after 6:00 p.m., and it became "a complete black box": shaded from the outside world so that the *Oryza japonica* plants would get only eleven and a half hours of daylight, as they would have in the Asian countryside. For experimental integrity, a computer randomized the seedlings' placement in the room, and among them were seedlings free from any climate-ready genes, controls there to suffer—presumably worse—alongside their enhanced counterparts.

We followed the conveyor belts to a tall box at the far corner of the greenhouse: the "imaging cabinet," or ARIS—Automatic Rice Imaging System—which was like an MRI machine for plants. Each seedling visited it once a week to be photographed from seven different angles, including through the walls of its tailor-made clear pot. The goal was to measure "vegetative parameters." In the images, total pixel area was a proxy for total biomass—"Data is extracted from the pixels. It's the pixels," Peferoen said, nodding excitedly—and root development was judged by the number and width of lines photographed underneath the plant. The seedlings raced through the machine at a blinding pace, eight hundred an hour, seven thousand a day, each illuminated for a few seconds by flashes of light, and were then cycled back into their artificial paddy. The images were 3 megabytes apiece, Peferoen said; approximately 50,000 pictures a day meant 150,000 megabytes of data a day. So they waited until Internet traffic was low to transmit it all to BASF's computers for analysis, sending it out overnight in batches as the rest of Belgium slept.

Ghent, Belgium, was one of the birthplaces of biotechnology, where in the early 1980s scientists learned how to transport genes into plants by infecting them with bacteria. By the time I visited this BASF subsidiary, CropDesign, the industry had grown so much that I had three media handlers from three different countries. The German, the American, and the Belgian had sat me for hours at a mustard yellow table in a room with mustard yellow walls and buffeted me with PowerPoint presentations. The German provided stats on GM crops: In 1997, the year before BASF, the "chemical company," had jumped into genetic modification, 25 million acres were planted worldwide. In 2011, 400 million. In 2020, herbicide tolerance—the trait that made the industry's fortunes up to now—would be worth less than 100 million euros. The traits that constituted what CropDesign called "intrinsic yield"—drought tolerance, salt tolerance, stress tolerance—would be worth 2 billion euros. The Belgian explained what Norman Borlaug had achieved in the 1960s with the Green Revolution: density. "Take the individual corn plant forty years ago and the individual corn plant today," he said, "and the difference is not that much. The big difference is that when I was a child,

I could run through the cornfields and build houses in the cornfields. Now there is no way you could get in between them." Monsanto had recently pledged to double corn, soybean, and cotton yields.

CropDesign's CEO came in to explain that what I was seeing was part of the company's trademarked TraitMill process, a "highway from gene selection to patent filing." Promising creations from the greenhouse were sent on to field trials in the United States or Brazil or to another BASF subsidiary in Germany, where they documented changes to amino acids with every tweaked gene. So far, 150,000 patents had been filed— one for every amino change. "We only file when we see validated data in the crop," he clarified. Already, they had identified traits for 50 percent higher yield, 30 percent larger seeds. The CEO's pride in the system shone through bursts of corporate-speak. "TraitMill is the largest validated crop-based platform for the development of productive traits," he said, beaming. "And it's IP-protected!"

THE TYPICAL FLIGHT RANGE of *Aedes aegypti*, genetically modified or not, is about a hundred meters, and the walk between Luke Alphey's office and Oxitec's mosquito nursery was at least twice that. I still hadn't seen an OX513A, so before he and I grabbed lunch, we took a stroll through the industrial park. Dengue comes in boom-and-bust cycles, he reminded me. "The only time anyone really notices," he said, "is like when *albopictus* showed up in the United States, or now dengue in Key West." The important thing was to keep our resolve during the inevitable lulls.

Inside the nursery, staffers and grad students in white lab coats huddled over microscopes, and in one room I peered through an eyepiece to see a modified *Aedes* larvae glowing red. "All of our constructs have these fluorescent markers," Alphey said, thanks to coral and jellyfish genes Oxitec had inserted. He pushed open a door, and we stepped into what looked like an oversize closet. It was a steady eighty-two degrees inside. Two dozen BugDorm-brand insect cages ringed the moldy walls. The overhead lamp emitted a distracting electric hum. To the side stood the unlucky employee who spent her whole day inside this humid, mosquito-filled room, which seemed a decent analogue for Key West or the Caymans but for the lack of sunsets. Here Oxitec was producing two million mosquitoes a week, Alphey said. He showed me a tray of water with dozens of larvae and a few that had already become pupae, which swam around looking like tiny, tiny tadpoles. "In the light, they all huddle in a corner," he said. "See? In the dark, they mellow out." He showed me a foot-long strip of paper that had perhaps forty thousand dried eggs. They kept for a long time—long enough to ship around the world. Add to water, put in a vacuum, get mosquitoes: "You can get a nice synchronous emergence." He showed me a small plastic cup barely filled with what he said were a million eggs. They looked like coffee grounds.

The adult OX513As clung to the walls of the BugDorms, hundreds per cage. When they flew, there was barely a sound. *Aedes aegypti* doesn't have the annoying buzz of other species, and it seemed appropriate that warmer temperatures might affect its range: Like climate change, people didn't really notice it until it was in their face—and then they tried all manner of crazy things to stop it. Below the cages, I noticed, was "the Executioner," a handheld bug zapper shaped like a tennis racket: insurance for whenever a mosquito made a break for it. "What we're doing in here is optimization," Alphey said. Oxitec wasn't shipping these eggs to the tropics, not yet. "We want to just try and improve the rearing process, or the cost gets unbearable. What's the best number of adults to have in the cage, and how long do you keep them in the cage, and how do you feed them, and when?" For now, Oxitec fed the mosquitoes fish food. "Like you give a goldfish," Alphey said, "but you can grow them on yeast powder, dog biscuits, cat food—whatever organic material happens to be in the water." For all the groundbreaking genetics, there was also this mundane side of the business, he said: "How do you make large numbers of inexpensive yet fit, healthy, sexy male mosquitoes?"

It was a question befitting our Anthropocene epoch. Godlike power was beginning to feel normal, even tedious. In the United States, genetically modified crops have penetrated the market almost completely since their arrival less than twenty years ago: They make up 94 percent of our planted cotton, 93 percent of our soybeans, 88 percent of our corn. They have spread to two dozen other countries, and the value of the global GM market has jumped by 7,500 percent. The numbers will only rise along with the temperatures, for the world is on the verge of seeing not only crops with drought-resistant tweaks but millions more farmers—Chinese, Nigerian, Indonesian, Brazilian—with just enough money to buy genetically modified seeds. It is Oxitec, not Monsanto, that may be the true harbinger: Scientists are warming up to the idea of modifying bacteria and wild animals—not just crops—to adapt to the new climatic reality. In 2012, a study by NYU professor S. Matthew Liao proposed reengineering humans themselves to produce smaller, less resource-hungry, less emissions-intensive offspring. Months later, the first conferences on using "de-extinction" and "synthetic biology" to preserve the natural world were convened by the National Geographic Society and Wildlife Conservation Society. The Sahel need not become the Sahara if we can create a GM bacteria that induces plants' roots to grow. The polar bear need not ever go extinct. We can already manipulate stem cells. We can already reconstruct lost genomes. We can already clone. If a species disappears because Arctic sea ice disappears, we already have the power to bring it back to life.

Compared with what the future could hold, Alphey's mosquitoes were straightforward. All the OX513As in the room came from a single forebear Alphey had created a decade ago. From then on, this line had not been a GM program so much as a breeding program. "When I say we make transgenics by injecting DNA into their anuses, people then think that every one of those millions has to be injected. And they think, that's never going to be economical—and they'd be right but wrong. You do that once." I tried to follow along as he excitedly explained the process. In 2002, after he'd made the synthetic DNA, a technician had lined up a bunch of tiny *Aedes* eggs, all facing in the same direction. "Then you inject them with this fancy laser needle," he said. "The cells that form at the posterior pole are the GM-line precursors. They're going to become the sperm and the eggs when this little egg turns into an adult. If you can get your DNA into one or more of those cells and it's taken up into chromosomes—which is a very low-efficiency process— then a proportion of the sperm or eggs that the adult produces will have your bit of DNA in it." As Alphey talked, a renegade mosquito alit on his neck, and he mindlessly swatted at it. It was old-fashioned but deadly effective.

TWELVE

PROBLEM SOLVED

OUR GEOENGINEERED FUTURE

N athan Myhrvold's new laboratory was unmarked and, from the outside, unremarkable: a 27,500-square-foot former Harley-Davidson service center in an industrial suburb of Seattle, near a plumbing supply distributor and the evangelical Blue Sky Church. Among the cars in its parking lot, I counted an equal number of Priuses and Mercedes-Benzes—three and three—and near its entrance I saw a growing pack of technology bloggers and local television crews, here for a ribbon cutting. We were allowed inside before Myhrvold arrived, and present already were some of his scientists and inventors in white lab coats, standing casually at their stations, spread out across a checkerboard floor. We could not yet see any of the lasers, and we could not see any of the mosquitoes that we understood would be shot down by the lasers. Nor could we see the solution to climate change, even if, rumor had it, it was being invented and patented here.

When Myhrvold showed up, he was flanked by Maria Cantwell, Washington State's junior senator. He was bearded and rumpled, all boyish smiles and gesticulating arms, while she had a studied calm. He wore loose-fitting khaki pants and a jacket but no tie, while she wore a black pantsuit. The conceit of the media event was that Myhrvold was giving Cantwell a private tour, and we in the media clustered around them, politely stepping out of view of one another's cameras when the need arose, so as not to ruin the effect. "This is Philip," Myhrvold said at the first station, introducing a young man in a lab coat. "He just got his Ph.D. at Princeton." Philip showed them software that modeled malaria epidemics in Madagascar. The research, Myhrvold explained, was partly underwritten by Bill Gates and the Gates Foundation—which would try almost anything to stamp out mosquito-borne malaria. "Bill is an investor in our company," he said. "This stuff is sort of pro bono, but some of it, I think, will have a very profitable spin-out—we'll do well by doing good."

At Microsoft, Myhrvold had been Gates's in-house futurist and chief technology officer. At Cambridge, he had been a theoretical physics researcher under Stephen Hawking. He was a subject of Malcolm Gladwell profiles, darling of TED talks, and author of a 2,438-page, fifty-two-pound "modernist" cookbook—a man both celebrated and feared in tech circles. The grand opening of this lab, those of us following the tour understood, was meant to be a retort to critics of his post-Microsoft business, a $5 billion investment firm called Intellectual Ventures (IV). The company was accused of being a "patent troll": quietly buying up patents without producing anything of its own, and using the patents to extract licensing fees from those who did produce things—Verizon, Intel, Nokia, Sony—anytime it decided its intellectual property rights were violated. Its business model, critics said, was to threaten to sue. At the time of the ribbon cutting, IV had twenty-seven thousand known patents, though outside consultants believed a higher number was hidden among more than a thousand affiliated shell companies. IV spent a million dollars a year lobbying against patent reform. But in articles and interviews, Myhrvold rejected the "troll" label: The company had earned more than a billion in royalties but at the time of the tour had yet to sue anyone. And the lab was proof that IV was creating its own patents—some "five hundred to six hundred a year," he said.

From Philip's computer bay, Myhrvold led Cantwell and the rest of us into a conference room where eleven chairs flanked a long table. Flat-screen monitors hung in all four corners of the room, all showing the same video: a mosquito flapping along in slow motion until it was hit by a laser beam and it spiraled out of view. "Here's where we get together with various scientists to brainstorm new ideas," Myhrvold said. "We bought the fancy table and chairs in a bankruptcy auction. We're trying to expand our activities, making these long-range bets, whereas the rest of the world is retrenching." In the next room, he described a kind of hyper-insulated cooler to keep vaccines refrigerated for months in places without consistent electricity. "It's like a Coke vending machine," he said. We then donned goggles and filed through a door marked with a caution sign: "Big Scary Laser. Do not look into beam with remaining eye." Inside IV was developing a method to test for malaria using lasers rather than blood work, another Gates project. In a nearby room was the insectary: a mosquito-filled closet much like the one I saw at Oxitec, BugDorms and all. "We grow our own," Myhrvold said to Cantwell. "If ever you need to convince some people in Congress to do the right thing, you just need to hold a meeting here and lock the door." A doctor in a lab coat was standing by. "She has a Ph.D. in mosquitoes," he said. He gestured behind her. "See, there's the raisins. Turns out we mostly feed them raisins."

With the cameras rolling, Myhrvold soon offered Cantwell a steaming ball of liquid-nitrogen-dipped citrus foam inspired by his cookbook. But the main event was the mosquito zapper: We huddled over a partial prototype, a camera zoom lens that captured the bugs mid-flight so their flight patterns, wing-beat frequencies, and speeds could be analyzed. The idea was to be able to distinguish enemies from innocents—mosquitoes from bumblebees, bloodsucking females from innocuous males. In the final design, a low-powered laser would do the targeting, tracking the bugs on a screen as if it were a video game, or the cold war, and a more powerful laser would take them out. Cantwell climbed up a ladder to watch mosquitoes being targeted inside a ten-gallon aquarium—a green flash every time it locked on. "It's hitting about every two seconds," she said. "It can do about fifty per minute," Myhrvold corrected her.

The zapper was known as the photonic fence, and its similarity to the Reagan-era Star Wars scheme—nuclear-powered X-ray lasers orbiting in space, zapping Soviet missiles—was no coincidence. It was a brainchild of the astrophysicist Lowell Wood, Myhrvold's longtime friend and an associate at IV, who had developed and led the Star Wars program at the Lawrence Livermore National Laboratory (LLNL). The tie-dye-wearing Wood was a protégé of the LLNL's co-founder Edward Teller, the "father of the hydrogen bomb" who tested his wares in the Marshall Islands and was an inspiration for the character Dr. Strangelove. Wood and Teller had been fellows at the Hoover Institution, the libertarian-conservative think tank housed at Stanford University.

In the 1990s, Wood and Teller were among the first to seriously study planet-scale engineering to reverse climate change, which has come to be known as geoengineering. Their idea, described in a paper submitted to the Twenty-second International Seminar on Planetary Emergencies, was to mimic volcanoes. Find a way to spray sulfur or other aerosols into the stratosphere, and it would be like the aftermath of 1991's Mount Pinatubo eruption: The particles would shade the sunlight, and the global temperature would drop. That Intellectual Ventures had begun patenting geoengineering technologies—methods to stop hurricanes, refreeze the Arctic, and engineer the climate back to "normal"—was mostly speculation at the time of the lab's grand opening. Myhrvold said nothing about it to Cantwell, but later that day, in a brief conversation watched over by two wary handlers, he told me the rumors were true.

The tour ended under a gray sky in the parking lot. Inside a white tent, champagne and salmon were being served to a crowd of dignitaries, including venture capitalists, University of Washington professors, and the celebrated wildlife photographer Art Wolfe. He was a friend of

Myhrvold's, himself a published photographer. Myhrvold and Cantwell stood in front of a red ribbon, and when the cameras were ready, he announced that scissors were "a boring way to cut a ribbon, so we've devised something else." A staffer wheeled up a stand with a big red detonator button on it. Two others stood by wearing fireproof gloves, holding fire extinguishers. "Is the device armed?" Myhrvold asked, and then he began counting down: "Five, four, three, two, one." Cantwell pushed the button, the ribbon burst into flame, and the audience clapped wildly.

CLIMATE CHANGE WAS a new excuse, but inventors have always wanted to do something about the weather. In July 1946, one of the researchers in the General Electric lab run by Irving Langmuir, the chemist and Nobel laureate, accidentally invented cloud seeding, the forebear of geoengineering. He dropped a piece of dry ice into a cloud chamber, and the cloud was instantly transformed into ice crystals. "Control of Weather," scrawled Langmuir in his notebook. One of the lab workers was Bernard Vonnegut, brother of Kurt. Bernard would discover and eventually patent the fact that silver iodide serves even better than dry ice as a seed crystal, pulling a cloud's humidity into flakes or drops until they fall out of the sky. In his 1963 novel, *Cat's Cradle*, Kurt, who worked in GE's public relations office, would describe ice-nine, a fictional seed crystal that turns liquid water solid. The Langmuir lab ran its first cloud-seeding trials in November 1946, dropping six pounds of ice pellets into a cloud over New York's Berkshire Mountains, seeming to create three-mile-long bands of snow while winning headlines across the world. Within five years, commercial rainmaking operations, most using silver iodide, covered nearly 10 percent of the United States. Walt Disney soon produced a comic strip, *Donald Duck, Master Rainmaker*, in which Donald flew a red prop plane into a cloud. "Goodnight!" he yells. "I overseeded!" Many scientists, particularly in the United States, now doubt cloud seeding has much efficacy, but modern rainmakers have worked in more than fifty countries, including Israel, India, Senegal, and Saudi Arabia.

Cloud seeding's allure was also its danger: Assuming it worked, we were now in control. After the Chernobyl disaster, according to the historian James Fleming in his book *Fixing the Sky*, Soviet authorities might have employed it to save Moscow from an approaching radioactive cloud. Silver-iodide-spewing bombers allegedly triggered rain over Belarus, where in some areas the rate of thyroid cancer among children would jump fiftyfold. More recently, China's Weather Modification Office set up its rocket launchers outside Beijing in advance of the 2008 Olympics, shooting up any approaching cloud before it could rain on the parades. Elsewhere in Asia, seeding's legacy was darker: During the Vietnam War, America's classified Operation Popeye weaponized the rain, working for five years to disrupt the seasons and lengthen the monsoon over the Ho Chi Minh Trail.

A year after Langmuir's big discovery, his lab was folded into Project Cirrus, a $750,000-a-year classified program led by the army, air force, and navy. It ran more than 250 experiments, including a campaign to suppress forest fires, between 1947 and 1952. The most dramatic was one of the first: In October 1947, researchers intercepted Hurricane King, a tropical storm that had just ripped past Key West and through southern Florida, as it headed back out into the Atlantic Ocean. They flew a bomber into the eye of the storm and dropped eighty pounds of dry ice. The hurricane did a U-turn and made landfall again near Savannah, Georgia, killing one person and causing $23 million in damages. That year, Langmuir paid a visit to Edward Teller in Los Alamos, where he bragged to the cold warrior about the damage done by his cloud seeding. Weather control, as he would tell the *New York Times*, "could be as powerful a war weapon as the atom bomb." It was the first known attempt at hurricane modification—which Myhrvold's Intellectual Ventures would one day tackle via other means.

I witnessed my first cloud-seeding operation in Australia during the Murray-Darling drought. In the Snowy Mountains, the reservoirs that gave hydroelectric power to Sydney had become depleted, and the region's privatized power company, Snowy Hydro, was desperate. The command center, in the city of Cooma, was a darkened room filled with computers and grad students. They tracked a passing cold front via an array of eight side-by-side monitors, waiting to radio the activation signal to their silver-iodide generators: thirteen solar-powered, treelike metal towers hidden throughout a protected wilderness.

In the lift line at the Snowies' biggest resort, Thredbo, Snowy Hydro had already put up banners touting its work on behalf of skiers: "Cloud Seeding This Winter to Improve Snowfalls." A manager told me how much he preferred cloud seeding to installing more slope-side snowmakers, which had been crucial on lower runs the more temperatures rose. Across Australia, the only complaint I heard was from a man in a dried-out farming town a hundred miles away, on the other side of the mountains. "The problem with seeding," he said, "is that you're deciding which region deserves the rain."

CASEY TEGREENE, Intellectual Ventures' top patent lawyer, ushered me into his office a few months after the ribbon cutting, and as the door closed, there was a flash of tie-dye in the hallway: a brightly clothed Lowell Wood, rushing by in his sandals. I'd wanted to hear how IV invented things before I heard more from Myhrvold about what things it had invented, and Tegreene had been volunteered to describe the standard method. A mountain climber, trail runner, and ultimate Frisbee player in his late forties, he coordinated what the company called "invention sessions": the process of sticking three to ten handpicked scientists, doctors, or engineers in a conference room for eight to sixteen hours and asking them to grapple with "big, interesting, well-stated problems." There was no metaphor that perfectly captured what went on

in the room, Tegreene said. "It looks sort of like brainstorming," he told
me. "It looks sort of like a physics or engineering class. It looks sort of
like arguing." I was reminded of scenario planning: all the smartest
guys, all in a room.

The topic of IV's first invention session, in 2003, was digital cameras.
There were now up to five sessions a month, tackling everything from
surgical techniques to metamaterials. "Let's say the topic was addressing
the Earth's albedo," Tegreene said. "Then the inventors we'd pick might
be physicists or material scientists, and most would be multidimen-
sional, polymathic types, like Lowell and Nathan." Three or four sessions a year had been dedicated to geoengineering—which had interested
Myhrvold almost from the moment IV was founded, Tegreene said—
and perhaps another ten a year touched at least tangentially on the topic.

The free-form sessions were typically held in the conference room I'd
seen at the lab, where video cameras and mics recorded every word and
transmitted the in utero inventions to Tegreene's sixty-three-person
patent group, which included nearly two dozen attorneys, most of them
with Ph.D.s of their own in aerospace engineering, computer science,
biochemistry, or mathematics. The inventors, who were flown in for a
day or two, were fed well—Indian food, Ezell's Famous Chicken, cook-
out barbecues in the early days—and they were paid modestly for their
time. IV also gave them a stake in any resulting patents. "But I wonder if
they'd do it if they didn't get any compensation at all," Tegreene said. "I
think they would. The people who like to invent with us, they like to
discuss interesting problems."

In general, he let the conversations wander where they might. "If they
start talking about how to invent a better buggy whip, we might steer
them away," he said. "The market for buggy whips has been down for
150 years. But sometimes the kernel of an idea, you're not even sure
where it comes from. Two people might start arguing about this or that
technique to make clouds precipitate things out. And suddenly you're
working on a better way to ionize the vapors in clouds."

There was a strange sound at Tegreene's window, which looked down on some trees and parking spaces near the Bellevue Club, an athletics facility. A flicker was attacking its own reflection. "He always sits there and pounds the left side of the window," Tegreene explained. He turned back toward me. "Sessions may migrate into new areas, and that's okay," he said. "We're not trying to develop next year's product. Typically, if you give a set of good problems to very bright people, it stimulates them to think of other ideas as well. If we figure out how to capture wave energy from water, for instance, that might lead to some valuable ideas."

A patent typically lasts twenty years, and IV's investors were expected to take the long view. Their money was reportedly locked up for more than a decade in its two funds, the $590 million, Asia-focused Invention Development Fund and the broader, $2.3 billion Invention Investment Fund II. Investors included such tech companies as Amazon, Apple, Intel, Microsoft, and Sony, which were also amassing patent war chests of their own, and the Rockefeller and William and Flora Hewlett foundations and the endowments of Brown, Stanford, Cornell, and the University of Texas, which managed money for future generations, not just today's. Big trends were more important than quick gains, and geoengineering was not out of place among the inventions IV was developing and patenting: advances in nanotechnology, semiconductors, nuclear energy, medical devices, and agriculture whose payoff might be many years down the road. In the meantime, the firm generated revenue— more than $2 billion by 2011—partly through patent-licensing agreements with tech companies, some of whom, confusingly, were its own investors. The agreements made lawsuits unnecessary, though the implicit threat of lawsuits was arguably what propelled the agreements.

IV would soon hold an invention session "about moving large things, like dirt and rocks," Tegreene told me. "How do I get to something that is under, around, below, or inside an object—or is very large and very far away from me?" Myhrvold had recently taken a helicopter tour of Canada's tar sands along with Bill Gates and Warren Buffett. They were guests of the $6-billion-a-year tar sands contractor Kiewit Corporation, which also had a hand in lining the All-American Canal. Myhrvold had noticed mounds of sulfur, a by-product of tar sands mining—and the main ingredient in geoengineering schemes then being patented by IV. "There were big yellow mountains of it, like a hundred meters high by a thousand meters wide!" he later told the authors of *SuperFreakonomics*. "And they stair-step them, like a Mexican pyramid. So you could put one little pumping facility up there, and with one corner of one of those sulfur mountains, you could solve the whole global warming problem for the Northern Hemisphere."

The official line at IV was that they were not pursuing geoengineering for profit. "Intellectual Ventures invents new technology as its main business, but we do not expect or intend that our climate technology inventions will make money," read an FAQ posted after its interest in geoengineering became publicly known. That morning in Tegreene's office, before the flicker attacked the window again, he described what happened to inventors' ideas once they left the free-form sessions in the conference room. "After a session, we do a process we call triage," he said. "We have a whole computer system to categorize the ideas. We have a series of four conference calls every week, with patent attorneys, business-development people, and support staff. We'll say, 'Okay, we're looking at geoengineering ideas. Here's idea number one. Is it better than the first idea already in the stack? No. Is it better than the second? No. Is it better than the third?' Okay, then we'll move it to number three in the stack, and then three moves down to four, four to five, and so on. We rank them."

What did "better" mean? "You're mapping a bunch of different factors," Tegreene said. "Whether or not it's gonna get good patent protection. Whether it's in a licensing-friendly industry. Does it have some commercial implications? Is there some broader concept that could be patented based upon it? Will it cost a lot in terms of technical support to file it? So it's a combination." Filing a patent is expensive, usually very time-consuming. IV's inventors spewed out thousands, perhaps tens of thousands of ideas a year, and most never made it to the top of the stack. Triage meant that usually only the most commercializable ideas got through. "If an idea is one of the top few in a certain area," Tegreene said, "then we start patent applications on it." Knowing all the steps involved, I still found myself wondering why they had gone to the trouble of filing for geoengineering patents.

ADVOCATES OF GEOENGINEERING, or at least of geoengineering research, tend to fall into three categories: scientists deeply afraid of runaway climate change; free-market advocates deeply afraid of mandated carbon cuts; and the capitalists or philanthrocapitalists who sustain them both. To find all three, I had only to visit two Washingtons—Washington, D.C., and Washington State.

In both Washingtons, flitting between conferences and meetings and panels and labs, were the scientists, notably Ken Caldeira of the Carnegie Institution at Stanford University, a celebrated climate modeler who had coined the term "ocean acidification." On the opposite side of the political spectrum from Edward Teller and Lowell Wood, he was dead set against geoengineering when the cold warriors proposed their Pinatubo option. Then he ran some numbers. More aware than almost anyone of what climate change would do to the world, he soon became a frequent visitor to Capitol Hill and a key IV inventor, though profit was apparently not his motive. In the case of a related climate patent, he promised to "donate 100% of my share of the proceeds to non-profit charities and NGOs."

Official acceptance of geoengineering was growing in step with global carbon emissions. After Barack Obama was elected came the first top-level scientific panels, starting with the Royal Society in the U.K. and the National Academy of Sciences in the United States. Then came geoengineering hearings in the U.K. House of Commons and the U.S.

House of Representatives, closed-door sessions led by DARPA (the Defense Advanced Projects Research Agency), studies by the Government Accountability Office and Congressional Research Service, policy statements by the American Meteorological Society and Britain's Met Office, a design competition by Britain's Institution of Mechanical Engineers, an ethics conference at Asilomar, a report by the Rand Corporation, a side event at the 2009 climate conference in Copenhagen, funding from the U.K. government for limited field research, a neologism-spawning endorsement ("climate remediation") from the Bipartisan Policy Center in Washington, and a place in the 2014 IPCC report. Across the papers and panels and symposia, the majority opinion was not that geoengineering should be deployed—just cautiously studied.

The most promising scheme was still the Pinatubo option, part of a set of ideas now known as solar radiation management, or SRM— another term coined by Caldeira. It was Benjamin Franklin who apparently first linked volcanoes to global climate. In 1783, when Franklin was stationed in Paris, a chain of Icelandic volcanoes had erupted for eight straight months, and temperatures in the Northern Hemisphere plummeted. "There existed a constant fog over all Europe, and great part of North America," Franklin wrote. "Hence the surface was early frozen. Hence the first snows remained on it unmelted. Hence the air was more chilled." Another promising SRM scheme came from the British professors John Latham and Stephen Salter, who later worked with Intellectual Ventures: unmanned, wind-powered yachts that would sail the high seas, seeding marine clouds with a spray of saltwater droplets, thereby raising their reflectivity, or albedo. A greater portion of the sunlight hitting the tops of these clouds would bounce back, and the planet would cool.

The second category of geoengineering advocate, free marketeers who often ignored scientists' careful distinction between research and deployment, I also found around Washington, D.C. "The underlying struggle between environmentalists and property rightists is really what's going on," one told me. He was a lawyer at a small Virginia think tank that sometimes veered into outright skepticism and would later sue for access to the prominent climatologist Michael Mann's e-mails, hoping to reveal climate science as a taxpayer-funded scam. "What we're fighting over is engineering the culture—that's mitigation—versus engineering the environment. That's geoengineering." Newt Gingrich, before his 2012 presidential run, echoed the sentiment. "Instead of imposing an estimated $1 trillion cost on the economy," he wrote in a letter to supporters as he tried to sink a climate bill in the Senate, "geoengineering holds forth the promise of addressing global warming concerns for just a few billion dollars a year. Instead of penalizing ordinary Americans, we would have an option to address global warming by rewarding scientific innovation. Our message should be: Bring on the American Ingenuity. Stop the green pig."

Gingrich was a senior fellow at the American Enterprise Institute (AEI), the éminence grise of conservative thought, home over the years to everyone from Milton Friedman to Dick Cheney. Outsiders still accuse it of climate-change denial. It has received funding from Exxon-Mobil, lobbied against the Kyoto Protocol, and offered scientists $10,000 for papers undermining the IPCC. But climate change was real, the co-director of AEI's geoengineering program told me in 2009. Now there were two questions: Do you want to do something? How much are you willing to pay? "There's a gap here," he said. "I don't see the American people likely to bear significant costs. The only answer is geoengineering." The other director had been hired after working most of a decade on market-based ways to cut carbon emissions. "I did as good a job as possible, and it fell flat," he said. "I became convinced that every economically rational plan will fall flat. Okay, so what follows? We're going to do a lot of adaptation. But what also follows is that adaptation is limited. So we're going to need grand-scale adaptation—geoengineering."

I was again witnessing the subtle shift in conservative thought: To fight over climate science was becoming less tenable. To fight over what to do about it was not.

In Washington State, Seattle had not only Intellectual Ventures but the University of Washington, reason for big names in the emerging field of geoengineering to visit for lectures and seminars, and it also had Bill Gates: a funding source. Through Nathan Myhrvold, Gates had met Lowell Wood, and through Wood, Ken Caldeira. Beginning in 2006, independent of his foundation but in keeping with its focus on techno-fixes, Gates received ad hoc tutoring sessions from Caldeira and another prominent geoengineering researcher. Beginning in 2007, after the pair lamented that there was little money for even the most basic geoengineering research, he provided some. His informal support eventually took on a formal name—the Fund for Innovative Climate and Energy Research, or FICER—and to date it has given out $5.1 million for assorted meetings and research projects.

Gates money paid for a turning point in recent geoengineering history. A series of private dinners at the margins of the 2008 American Geophysical Union conference, attended by Caldeira, Wood, one of the AEI co-directors, and more than a dozen senior scientists, was "the moment when the conversation moved from 'Can we do it?' and 'Should we do it?' to the much more focused '*How* do we do it?'" writes the journalist Jeff Goodell in his book about geoengineering, *How to Cool the Planet*.

That year, Caldeira and other top scientists also tackled the "how?" question at a workshop convened by the Novim Group, a new nonprofit modeled on the work of the JASONs, the informal club of scientific elites that has solved classified problems for various branches of the U.S. government—DARPA, the navy, the CIA—since 1960. Many of the people in the room actually were JASONs, and the study group was led by the JASONs' onetime head, the physicist Steve Koonin, then the chief scientist of BP, soon to be the undersecretary of energy for science under

President Obama. "Imagine the president calls you up and says there's a climate emergency," Koonin told the group. "How quickly can you respond? What do you do?" These JASONs, whose club was named after the Jason of Greek mythology, were again being asked to save the world. In a phone call in 2009, Novim's executive director mentioned that he had "just been invited to a meeting of high-net-worth individuals next week who are interested in investing in the area of geoengineering." He didn't want to share any names. "But you'd recognize some of them," he said. I later noticed that a Novim study on the global temperature record received $100,000 from Bill Gates's FICER—perhaps a coincidence, perhaps not.

In time, Gates would also give $150,000 to a University of Leeds professor who would analyze the clouds, $300,000 to a Bay Area inventor and entrepreneur who would lab test the feasibility of seawater sprayers for Latham and Salter's automated cloud ships, and $100,000 to the first systematic study comparing various ways sulfur and other aerosols could be launched into the stratosphere. Written by the drone manufacturer Aurora Flight Sciences, it investigated delivery methods including rockets, dirigibles, Gulfstream jets, suspended pipes, and the Mark 7, a sixteen-inch naval gun used on American battleships. Among the cheapest options was the Boeing 747, but the jetliner's top altitude may not be quite high enough for SRM. It could require a new model of airplane. Soon, there was another Seattle name in various geoengineering panels and reports: Boeing. The company was represented by its chief scientist and the vice president of its Illinois- and California-based Phantom Works, a defense and space unit that seeks, among other things, "to address potential new markets."

MYHRVOLD'S OFFICE WAS in a beige building in an office park about half a mile from Casey Tegreene's office-park office, about three miles from the Intellectual Ventures lab. If unglamorous, it was at least spacious. To reach it, one walked past a reception desk, past a beautiful photograph he took of a calving Patagonian glacier, past his collection of nearly a hundred vintage typewriters, and past what appeared to be the skeleton of an allosaurus. (As a hobby, Myhrvold hunts dinosaur bones with the famed paleontologist Jack Horner; his money and drive have helped expand the world supply of *T. rex* specimens by 50 percent.) Inside the office itself was a cast of a prehistoric fish head about the size of a Smart car, along with a photograph of Myhrvold on a fly-fishing trip to Oregon's Umpqua River. It captured him grinning maniacally while pointing at a six-inch rainbow trout he'd just caught—surely the smallest fish in the river that day. When I walked in, he was sitting at a wooden desk, surrounded by three computer screens, cradling a Coke Zero. His shirt was half tucked in, and he was wearing socks with his Teva sandals.

Myhrvold was just beginning to talk publicly about his company's geoengineering inventions. "The reason this stuff works is interesting," he began. "The sun radiates an average 340 watts per square meter on Earth. What's called radiative forcing—which is the amount of extra heat trapped by CO_2—is today 2-something watts per square meter, and if it doubles, it'll be about 3.7 watts per square meter. That's roughly 1 percent of the energy from the sun! So a very crude way to think about this is that global warming is the accumulation of that 1 percent, like a penny on every dollar." A crude way to think about SRM, he suggested, was that we were returning that penny: "If you make the light 1 percent dimmer, you're there!"

The choice of sulfur aerosols as a dimming agent was somewhat arbitrary: While nanoparticles or tiny mirrors might also do the job, sulfur seemed safer because it was precisely what volcanoes spewed, and because it was already very present in nature. "It's natural and it's been around for literally billions of years, so to some extent what you see is what you get," he said. Pinatubo and other volcanoes had provided proof of the basic concept, so the main issue, in IV's view, was how to get aerosols high into the atmosphere in absence of an eruption. "We wanted something that we thought was more practical than the schemes we'd seen before," Myhrvold continued. "Is there a clever and cheap way to deliver the stuff to the stratosphere?"

When Lowell Wood retired from Lawrence Livermore in 2006 and moved north to work with Myhrvold, existing ideas for delivery—artillery, jet fuel doped with sulfur, and so on—had "a certain Rube Goldberg quality," Myhrvold said. "Now, some would say that me saying 'Rube Goldbergish' is sort of like the pot calling the kettle black, but anyway . . . imagine thousands of cannons pointing straight up, firing every day, all day. It's kind of a crazy scenario. And it's expensive—billions of dollars a year. Now, billions of dollars is still really, really cheap compared to many of the other things that people would compare it to. Suppose global warming happens without us making much of an intervention, how much are crops going to be ruined, how much will the economy be hurt? And we'll have to do things to try to cope. One example is that cities on the seafront—say, the Italians in Venice—will have to build seawalls or move. That's really, really expensive." In a series of invention sessions, IV came up with two new methods to do SRM. "Well, then we got on a roll and came up with other methods using other kinds of geoengineering," Myhrvold said, "but just for radiation management, we came up with what we believe are the most practical systems that anyone's proposed to date."

The first method was to pump the sulfur up into the stratosphere through a hose supported by a series of balloons: the "string of pearls." "I came up with the name," Myhrvold said. "The second method, which is actually the first, but we think of it as the second because the other one is better, is to make twenty-five-kilometer-tall inflatable smokestacks to take coal plant emissions and deposit them in the stratosphere." When emitted at a lower altitude, sulfur dioxide, a major by-product of coal burning, causes acid rain; because of it, coal plants in the United States have been highly regulated by the Clean Air Act since the 1970s.

The idea would seem to make expensive sulfur scrubbers obsolete. "Lowell Wood came up with the inflatable smokestack," Myhrvold said, "which he started explaining to us as a 'toroidal balloon.' At first we didn't understand. Technically, it is a torus—a doughnut—but because one axis is stretched twenty-five kilometers, thinking of it as a doughnut is just weird. Your head has to be on the wrong way—but Lowell's is! He's a very creative thinker." Hot air rises, the smokestack could be insulated, and all the math seemed to work out. But there were many uncertainties with the method. "For instance, nobody has ever made an inflatable smokestack twenty-five kilometers high," Myhrvold said. The inventors moved on.

"It was like, why don't we just run a hose up there and pump it?" Myhrvold said. "But it's difficult because of the hydraulic head, so it was like, screw that, let's just have a whole lot of pumps. If you have pumps every hundred meters, it's really simple." Two of Myhrvold's employees had recently won $900,000 at NASA's Space Elevator Games. "Their laser-powered robot climbed a 900-meter-long cable suspended from a hovering helicopter in less than 7.5 minutes," read the press release.

"If you're doing a space elevator," he said, "you know totally the following thing: The longer the rope, the stronger it has to be. Any rope, if you make it long enough, will break from its own weight." While it might be technically possible to use a single pump and a single blimp for SRM, the string-of-pearls approach seemed far superior. "You can support it all the way along," he said, "then the structural problems of a very long hose go away." And unlike the twenty-five-kilometer smokestack, all of the components existed already, though the spray mechanisms would need to be improved.

After the original, toroids-and-pearls discussion, IV's team—usually Wood, Myhrvold, Caldeira, Tegreene, and various others—refined their ideas in half a dozen more invention sessions. When they finally went public, releasing an eighteen-page research paper filled with futuristic images, they would dub their invention the Stratospheric Shield, or

StratoShield for short. They proposed that early efforts could focus on the Arctic, where temperatures were shooting up fastest and thinning ice was leaving a planetary bald spot—the yarmulke method, as it was known in geoengineering circles. In order to reverse worldwide warming from a doubling of CO_2, climate modeling suggested that 2 to 5 million metric tons of sulfur dioxide would need to be pumped into the stratosphere every year. But a rough estimate for just the Arctic was 200,000 tons. IV envisioned several 100,000-ton-a-year, 7-ton-a-minute pumping stations scattered across the region, operating only in the spring—because during winter the Arctic was already dark. Hoses would deliver liquid sulfur dioxide to an altitude of approximately twenty miles, where a series of atomizers would spray out a mist of hundred-nanometer aerosol particles. Average temperatures would drop five degrees Fahrenheit, the paper said, and sea ice would go back to its preindustrial extent. The rough price tag per pumping station: $24 million, including transportation and assembly, plus $10 million in annual operating costs. That is, when compared even with a single flood barrier for New York City or seawall for Seattle, effectively free.

I pointed out that the invention would do little for ocean acidification, and Myhrvold readily agreed. "Yeah, but I think we have a solution for that, too," he said. "First of all, ocean acidification, that whole phenomenon, was first put into the literature by Ken Caldeira, who works here. But before I get into that, I should tell you about our hurricane suppressor." At an early geoengineering meeting at Stanford put on by Caldeira and Wood, Myhrvold explained, Stephen Salter had shown up, and soon he was recruited to work with IV on projects including his cloud-whitening concept (for which he, not IV, had the patent). "But he had another brilliant idea," Myhrvold said, "so we started improving it, and now we have this very cool way to reduce the strength of hurricanes."

The Salter Sink, like other hurricane-suppression schemes, including that of the New Mexico company Atmocean, was designed around the fact that hurricanes derive their energy from the heat of the ocean. Higher surface temperatures, as was the case before Sandy, and you get bigger storms. Lower surface temperatures, smaller storms. "This would be a useful thing even without global warming," Myhrvold said. "But it's very likely that these storms will be stronger due to global warming." IV's idea was to pump warm surface water down to the colder depths, thus cooling off the top layer—a mechanized churning of the sea. The sinks themselves were large, floating rings, up to three hundred feet across and made of used tires, with attached tubes—which they called "drains"—stretching hundreds of feet down. Deploy seven hundred Salter Sinks in the path of a category 4 hurricane in the Gulf of Mexico, IV's research suggested, and the storm would effectively disappear.

In one invention session, Wood had the epiphany that the same churning process could be applied to ocean acidification: The acid concentrations that mattered were those at the top of the water column, where the majority of sea life resides. "So we think it is possible to tackle ocean acidification," Myhrvold said. "If we put a bunch of these Salter Sinks in, then we'll turn the surface over, and if we turn the surface over, we effectively dilute any acidification that may occur. This approach is not 100 percent proven yet, but Ken and some collaborators have some modeling going on."

In late 2009, not long before the Copenhagen climate conference, IV released a paper on hurricane suppression. It contained the company line on geoengineering: This research was for the good of the world, not for the good of IV's investors. "As with its other geoengineering inventions such as the Stratospheric Shield," it stated, "Intellectual Ventures does not advocate immediate construction or deployment of Salter Sinks. Indeed, IV sees no immediate business model to support development of this technology. Our hope in publicizing the invention is to suggest that practical defenses against at least some catastrophic storms may be possible."

I later saw the patent applications for hurricane suppression, which bore Myhrvold's name, along with those of Gates, Salter, Latham, Wood, Caldeira, Tegreene, and various others. The business model might not have been "immediate," but it existed all the same: In addition to describing the mechanics of the Salter Sink, the documents described how a theoretical hurricane-suppression company could sell individual insurance policies. In one patent-pending scenario, the "ecological alteration equipment" would be moved into position on demand—provided there was "at least one payment from . . . at least one interested party." In another, the hurricane-suppression company would drum up potential clients by "alerting at least one interested party as to the potential for storm damage . . . providing information to at least one interested party of the cost and likelihood of reducing damage . . . and receiving at least one payment." IV was attempting to patent a new insurance policy for the global-warming age—Firebreak's basic business model, only applied to hurricanes, not wildfires.

MYHRVOLD AND IV's geoengineering ideas were first introduced to the world in the pages of *SuperFreakonomics*. The authors Steven Levitt and Stephen Dubner's take on climate science and support for geoengineering as an alternative to emissions cuts seemed to suggest they had spoken to few scientists but those at IV, and criticism of the book was intense. Myhrvold, too, was drawn in, and he felt burned.

"Some climate activists take the position that we should forestall any debate about a broad set of solutions," Myhrvold said at our last meeting. "They have *the* one solution—which is to cut back and go renewable and so forth. They hate the idea of geoengineering to death. They have an ideology of conservation, of living lightly, that is in some cases very antitechnology. And if you have that ideology, then global warming is finally the justification to convince people of what you want." He believed he understood why geoengineering upset them. "They say, if there's an easy way out, people will take it," he said. "Now, my reply is:

It's not like you guys have made any progress whatsoever. Zero. Zip. Nada. Some tax dollars have been wasted in Germany and the United States subsidizing noneconomic things. The idea of Germany as a solar energy hub is just ludicrous—and it's very likely that those German solar installations cause net harm to global warming, very likely. I haven't done all the calculations, but it takes a lot of energy to make solar plants, and if it's cloudy all day, you don't get much benefit back out." (The IPCC, on the other hand, had begun some calculations, and photovoltaic energy produces roughly twenty times fewer life-cycle greenhouse emissions than natural gas, forty times fewer than coal.)

For a moment, Myhrvold seemed to channel the American Enterprise Institute, questioning whether money spent cutting emissions was being put to the best use. The emissions-only approach was "particularly rude to the poor people on Earth," he said. "We're a wealthy country, so we could afford to do a bunch of stuff. Poor people can't—or they won't. In Asia, in China, they want their industrial development, and I don't know how you can stop them." In Africa, the logic was starker. "Those people live right on the edge," he explained. "Now there are people who say: 'Ah, climate change will hit them worse.' True. But if you're starving to death already, or you're dying of malaria or a bunch of other diseases that could be alleviated with tiny amounts of money compared with what the rich world might spend on climate change, there's an interesting moral issue: How much should we spend?"

He offered what he said was a "deeply politically incorrect" analogy for climate politics. "This is very much like what happened when some religious groups found HIV/AIDS," he said. "It was like, 'Look, we've wanted people to not be gay, to not have homosexual sex for a long time. Here is a heaven-sent opportunity—God's punishment against gay people or drug users or the promiscuous.'" Some environmentalists' ideological response to geoengineering proposals, he said, was "very much like the pope's position on HIV. The pope said that condoms are not the answer. Well, with all due respect to his infallibility in matters of faith, empirically, he's just flat out wrong. Preaching abstinence as a remedy for HIV is ineffective. The fact is abstinence doesn't work. We all do things that we know are in our own long-term worst interest."

If people "won't stop having unsafe sex" and "won't stop eating ourselves to death," he wondered, how could they tackle climate emissions? "I'm afraid that preaching abstinence with energy is like preaching abstinence with Krispy Kremes or abstinence with sex," he continued. "It's a message that says, 'Yes, forty years from now or a hundred years from now, in 2100, things will be really bad—that's why you shouldn't use your energy today.' If people won't get that if they have unprotected sex today and it'll kill you in a few years, why should this other message get across any better? It's morally bankrupt for the pope to say abstinence only for people to fight HIV—it kills people to say that—and I think that's a worse sin than fucking.

"I don't think the world is ready to do anything about climate change," he said. "I could be wrong, but I say that right now you can sort the world's countries into two categories: Countries that say climate change is a top priority yet have done absolutely nothing. And countries that say, fuck it, we're not doing anything. So what have we done? Nothing! Zero! Europe has a little bit of carbon trading, but nobody will tell you it's done anything—it's all been window-dressing bullshit. Now, you tell me, where's the optimism?" He took a swig of his Coke Zero. "We got interested in geoengineering," he said, "because once all is said and done, more is said than done."

Without prompting, Myhrvold answered the question I had after the meeting with Tegreene: If IV was researching geoengineering for the good of the world, why the patents? "It's a little bit of a crazy area," he began, "because we invent for, uh, money. We're a company; we're a for-profit organization. And it's very hard to see how you actually make money from any of these schemes. Someone could go do it in some other country, and it's unclear we could go ask them for anything for our patent. It's not like this is something you sell a million copies of. We decided to actually file patents in part because we file patents—it's what we do—and in part because we thought this may give us a seat at the table in deciding whether and when the technology is used." It sounded perfectly reasonable. I still didn't know whether to believe him.

In time, IV would suffer another media controversy. A report by the radio program *This American Life* exposed how a company that made a point of telling journalists that it didn't file lawsuits apparently saw to it that its patents were used to file lawsuits. It sold them to affiliated companies, the program reported, on the condition that it get a large share of any profits they might earn—and it was the shell companies that sued. The reporters visited an IV-affiliated shell company, Oasis Research, in twenty-four-thousand-person Marshall, Texas, finding an empty office in a two-story building filled with other empty offices also involved in patent lawsuits, and slowly the back-end arrangement was uncovered. The report was incriminating but perhaps liberating, too. By 2011, IV was filing patent-infringement suits openly, under its own name, trying to shake money from no lesser targets than Motorola, Symantec, Dell, and Hewlett-Packard.

The geoengineering invention sessions eventually stopped; IV had dreamed up what it could, the patents were pending, and now it seemed to be up to the world what happened next. "In general, global warming is about the worst possible problem for our society to deal with," Myhrvold told me. "Almost the worst-case problem for our psychology. We deal best with ecological problems that are severe and localized in space and time. When you have something like that or the *Exxon Valdez* spill or Love Canal, when the impact is immediate, local, and extreme, then it's pretty easy. This is also true of forest fires. It's like, 'Oh, God, we've got to fix this!'" He paused. "The problem with global warming," he said, "is that it's not localized in space—it's global. And it's not localized in time, either. We're just not set up to deal with it." Myhrvold didn't need to make the case that geoengineering was desirable. He made it seem inevitable; we wouldn't manage to do anything else.

I felt myself succumbing to the logic. It didn't matter, I realized, whether or not IV was secretly hoping to turn a profit—whether or not its assertions that it did not "expect or intend" to make money off geoengineering were as misleading as its assertions about patent lawsuits. If one of Myhrvold's inventions could save the planet, it was beside the point that he might get richer in the process.

I DID NOT fully understand what might be saved by geoengineering until one January afternoon in Seattle, on a day that was warmer than average, but not abnormally so, and rainier than average, but not abnormally so, and a climate scientist named Alan Robock led a seminar at the University of Washington. A bald and bearded Rutgers professor with a brooding look and one too many buttons unbuttoned on a striped shirt, Robock focused on a geoengineering problem that went largely unmentioned by AEI, unmentioned by Myhrvold: Fix the temperature, and you don't necessarily fix the rain. Use SRM to reverse global warming and get back to "normal," and you cannot be sure that precipitation patterns will follow course. Supercomputer climate models suggested that there would be a trade-off: If geoengineers wanted a certain temperature, they might twist the planetary thermostat in one direction; in some cases, for some regions, if they wanted a certain amount of rainfall, they might twist the dial further, or twist it in the other direction.

Geoengineering advocates claimed that volcanoes were proof of concept, Robock said—that they showed that there could be cooling from sulfur aerosols and that it was relatively innocuous. "But I've studied volcanoes my entire career," he said, "and I can tell you that they're not innocuous." Pinatubo had reduced rainfall in the Amazon and disrupted the Indian and African monsoons, according to a 2007 study by the National Center for Atmospheric Research, leading to local droughts. Later research by the Met Office in the U.K. showed that between 1900 and 2010, three of the four driest summers in Africa's Sahel had followed on the heels of a major volcanic eruption in the Northern Hemisphere. In the seminar, Robock presented anecdotal evidence that after the 1783 Icelandic eruptions that had so impressed Benjamin Franklin, collapsing monsoons might have led to drought and famine in India, China, and, most dramatically, Egypt, where a sixth of the country's population either died or fled within two years as the Nile ran dry. "Soon after the end of November, the famine carried off, at Cairo, nearly as many as the plague," wrote a French visitor at the time. "The streets, which before were full of beggars, now afforded not a single one: all had perished or deserted the city."

Robock flipped through his presentation, landing on a map of the world. The slide showed what resulted from running a version of the yarmulke method—an Arctic-only shield of sulfur not entirely unlike the one proposed by Intellectual Ventures—through his supercomputer models. It was just a model run, he cautioned, just one possible scenario. But as a vision of the future, it was instructive. SRM appeared to create a belt of abnormal precipitation patterns in the poorest parts of the world. According to the model, deploying something like a StratoShield would cut rainfall in the South Pacific, drying up island nations that might otherwise be drowned by sea-level rise. It would destroy the Asian monsoon, dumping extra water on Bangladesh, saddling India with permanent drought. It would destroy the African monsoon, turning Senegal and much of the Sahel into a band of brown, achieving through carbon emissions plus geoengineering what might otherwise be achieved through emissions alone.

But SRM, in this model, also promised to restore preindustrial temperatures and rainfall in most of North America, most of Europe, most of Russia, most of South America, and most of Australia. It would even improve the sunsets, Robock admitted. My eyes locked on Seattle and the American West: my home, my wife's home, our families' homes. Home to Bill Gates, his wife, his kids. Home to Nathan Myhrvold, his wife, his kids. It appeared that our corner of the planet could look just as it always has: normal temperature, normal precipitation. It's true that it rains a lot around here, but in the summertime the sun comes out. Everything is green. It's shockingly green. There are mountains to the east and mountains to the west, and there's water in every direction. In the summertime, there's nowhere in the world I would rather be.

I could guess which way the two Washingtons' geoengineers would twist the dial. It was then that I knew for sure that everything, for some of us, would be just fine.

MAGICAL THINKING

The year 2012, which ended as I sat down to write this epilogue, bound to forever be in my mind as the year my son was born, was otherwise apocalyptic. There was a tornado that hit Michigan on a seventy-five-degree day in March—I rode my bike to see its wake, passing prematurely blooming flowers that would die in the next freeze—and a drought that engulfed 61 percent of the United States, causing food prices to soar and saddling taxpayers with a $16 billion crop insurance bill. The Mississippi River ran to such record lows that barges either carried lighter loads or risked running aground. The Rockies got so dry that two states, Colorado and New Mexico, had the worst fire seasons in memory, and flames lingered in the mountains at ten thousand feet, where there would normally be snow. In thirty-eight states, especially drought-addled Texas, a record outbreak of mosquito-borne West Nile virus infected more than 1,118 people and killed 41. In Miami-Dade County, a woman contracted the first locally acquired case of dengue fever.

A third of America's population lived through at least ten 100-degree days in 2012. At weather stations across the country, there were 362 all-time record highs. There were zero all-time record lows. There were 2,559 monthly record highs—higher local temperatures recorded in January or June or November than in any previous January or June or November. There were 194 monthly record lows. There were 34,008 daily record highs—higher local temperatures recorded on an April 19 or August 24 or December 14 than on any previous April 19 or August 24 or December 14. There were 6,664 daily record lows. Grand Rapids, Galveston, Greenville, Albany, Billings, Boston, Madison, Nashville, Louisville, Chicago, Trenton, Richmond, and hundreds of other cities had hotter days than ever before. For the contiguous United States as a whole, the average temperature in 2012 was 55.3 degrees, 3.2 degrees warmer than the average day in the twentieth century. The average annual temperature was a full degree warmer than the previous record.

In the Arctic, 97 percent of Greenland's ice sheet was observed melting on a single day. Sea ice shrank to a dramatic new low, surpassing the record 2007 melt by another 300,000 square miles, another Texas. A record forty-seven cargo ships traversed the Northern Sea Route—a twelvefold increase since 2010. A floating, 644-foot-long gated community, billed as "the largest privately owned residential yacht on earth," transited the once treacherous Northwest Passage. Another yacht was soon impounded after its Australian owner served alcohol to a fifteen-year-old Nunavut girl, who dove partly clothed into the ice-free Beaufort Sea, and Mounties seized from it $40,000 worth of liquor and $15,000 worth of illegal fireworks. Global emissions had just jumped 3.1 percent, the price of carbon in the EU Emissions Trading Scheme was about to hit a record low below 5 euros a ton, the concentration of carbon in the atmosphere was soon to hit a milestone four hundred parts per million, and another round of aimless UN climate talks was slated for Qatar, which has the highest per capita emissions in the world—but it was a party.

Then, Sandy: the $60 billion hurricane that flooded the mid-Atlantic and gave New Yorkers a hunger for seawalls, gave President Obama's reelection a Bloomberg endorsement and arguably an electoral landslide, and gave climate change a prominent place in Obama's second term. Where Deutsche Bank's jungle tent once stood at South Street Seaport, there was a twelve-foot storm surge, followed by a stubborn rebuilding. Again we believe in climate change. Which raises a question: So what?

In psychology, magical thinking is the fallacy that thoughts correspond to actions—that to think is to do, to believe is to act. Perhaps the most magical assumption of the moment is that our growing belief in climate change will lead to a real effort to stop it. But as I discovered in Canada and Greenland and Sudan and Seattle and all over the globe, that is not automatically true. We are noticing that in this new world, there is new oil to find. There is new cropland to farm. There are new machines to be built. From what I have seen in six years of reporting this book, the climate is changing faster than we are.

I SPENT PART of the summer of 2012 in the Inupiat village of Point Hope, alongside the Chukchi Sea. It is an old place, older than almost any other on the continent. Its mayor, accurately or not, likes to say it's the oldest continuously inhabited settlement in North America. In his black Ford SUV, he drives visitors to a sandy spit being washed away by relentless waves, to a partly caved-in sod house propped up by whalebones, where his grandmother once lived. The first Europeans appeared in Point Hope in the 1840s, he explains, to hunt the bowhead whales once so numerous that elders claimed you could skip from one to the next, spout to spout, across the shallow Chukchi. Whalers were the original oilmen, magicians of a sort, for they rendered the bowheads into something the Inupiats had scarcely imagined: fuel. The first who came recorded the vitality of Point Hope in their journals—the many women, dwellings, and dogs they spotted from their ships. But in a few short years, the whalers' harpoons had killed most of the bowheads and thereby killed most of the people who depended on them. The homes were emptied and the few survivors gaunt, and all the dogs were gone; they had been eaten.

A century later, Edward Teller, fresh from nuclear-test triumph in the

Marshall Islands, still years from proposing to geoengineer the climate, arrived in Point Hope. He was determined to find a civilian purpose for the atomic bomb, and he had decided that a site twenty miles south of the village could be turned into the Alaskan Arctic's first deepwater port with a series of nuclear blasts. He called it Project Chariot. Other scientists soon calculated that the village would be decimated by fallout. But only after Point Hope became the focus of a national environmental campaign could the project's momentum be stopped.

Now the mayor was awaiting Royal Dutch Shell's refurbished armada of drill ships. A dozen of the most expensive oil blocks in Lease Sale 193 were offshore in the ice-free Chukchi, and the Obama administration, in its "all of the above" efforts to ramp up domestic energy production, had finally given Shell a green light. The company had by then spent $4.5 billion on Arctic leases and infrastructure, and the project seemed unstoppable.

When the *Kulluk* and another drill ship left Seattle's Puget Sound for Alaska after a series of upgrades, I had watched them go, the former's towering hulk towed by a massive, purpose-built tug, both rigs escorted by a security detail of coast guard vessels. There would be nearly twenty ships in the Shell fleet, three layers of spill-response capability, and thousands of workers and hangars and aircraft stationed across northern Alaska. I had my concerns about what Shell was doing, but I never doubted that a company so adept at planning—as collectively brilliant as was Teller individually—could pull it off.

Less than six months later, in the very last hours of 2012, the flagship Arctic rig of the world's most future-oriented oil company was beached on a rocky shoreline near Alaska's Kodiak Island. Its photograph was soon on the front page of newspapers around the world. The *Kulluk* had been on its way back to Seattle after a season riddled with missteps—dragged anchors, failed sea trials, EPA infractions—but had nevertheless drilled the start of the American Arctic's first offshore wells since the melt began in earnest. Its Louisiana-based crew towed it into a

960-millibar cyclone that Kodiak residents told me would have sent experienced Alaskan captains scurrying for safe harbor. The towline snapped in forty-foot seas, and over the next four days, through attempted rescues and emergency tows, it snapped four more times until the massive ship was finally grounded, waves slamming over its deck. Shell's $4.5 billion Arctic gamble was on the rocks, and having been honestly impressed with the company's brainpower, I was as shocked as anyone.

We are always wowed by the smartest guys in the room—the Tellers, the Myhrvolds, the experts, and the engineers—when we are in the room. As the world changes into an environment at least as foreign to many of us as Alaska was to a Louisiana tug captain, some of our smartest are developing staggeringly complicated plans to deal with what is essentially a problem of basic physics: Add carbon, get heat. We should remember that there is also genius in simplicity. We should remember that we rarely recognize hubris until it is too late.

THE SUMMER OF 2012 was hot in Seattle, too. Jenny and the new baby and I often slept downstairs, because upstairs was too warm and we'd never had much need for air-conditioning before. We went swimming more than usual; it was nice. We bought a bigger car—a relative gas-guzzler, but it fits the whole family. Our house is near Seattle's new light-rail, and during a remodel I made sure we insulated it well and got a high-efficiency furnace. But we drove all over that summer, and we bought a lot of Shell gas. The many flights I took, from a carbon perspective, were even worse.

One afternoon, after I watched the *Kulluk* get towed north, I went to the "prediction market" Web site Intrade and put a $100 wager down in its Climate and Weather category. I could have bet on global temperature anomalies or on Sandy being the last named storm of the 2012 hurricane season, but instead I chose the melting polar ice cap: "Arctic sea ice extent for Sep 2012 to be less than 3.7 million square kilometres." It was all in good fun, just a stunt to prove that any one of us, especially those in a comfortable place, could idly bet on climate chaos. But I won handily.

There is something crass about profiting off disaster, certainly, but there is nothing fundamentally wrong with it. I did not write this book to take aim at honest businessmen like Mark Fulton, Phil Heilberg, and Luke Alphey or at good soldiers like Sergeant Strong and Minik Kleist. If they are vilified because readers have not fully grappled with the landscape in which they live, the landscape in which we all live, then I have failed to describe it well enough.

The hardest truth about climate change is that it is not equally bad for everyone. Some people—the rich, the northern—will find ways to thrive while others cannot, and many people will wall themselves off from the worst effects of warming while others remain on the wrong side. The problem with our profiting off this disaster is not that it is morally bankrupt to do so but that climate change, unlike some other disasters, is man-made. The people most responsible for historic green-house emissions are also the most likely to succeed in this new reality and the least likely to feel a mortal threat from continued warming. The imbalance between rich and north and poor and south—inherited from history and geography, accelerated by warming—is becoming even more entrenched.

Environmental campaigners shy away from the fact that some people will see upsides to climate change—more minerals for miners, more famines for food sellers—because any local gains muddy the otherwise catastrophic picture of a world without emissions cuts. I have not shied away, for the people described in these pages reveal something important: In an unfair world, rational self-interest is not always what we wish it would be. In economic terms, global warming is not merely an externality that we have failed to price in. The free market can only get us so far. This makes it a truly wicked problem, but it also gives us a more perfect moral clarity. We are not simply borrowing against our own future. For the most part, we are not our own victims. To rely on empathy to shape our response to climate change is often considered naive—the victims of warming are distant in space, distant in time, and the bullets are invisible—but I believe it is more naive to hope that we in the north will significantly cut emissions or consumption or give needed adaptation funding to distant countries because we personally feel threatened.

In the world ahead, the politics of anger are not likely to work without the corresponding empathy. It is not enough to get mad at the oil companies—though it might help a little bit. There have been various postmortems about why the U.S. Senate has not passed a climate bill, or why the UN cannot get a treaty, but the reason is fairly straightforward: In the wealthy north, where we still talk more about polar bears than about people, there is no true constituency. Hardly anyone cares that much. Not yet.

When I was halfway through this project, I was checking facts with a source, an investment banker in New York who had acquired some foreign farmland. We got into an argument. What had happened along the way to his getting his tracts—a series of swindles by middlemen, of small farmers bought out by forces much larger than they could imagine—was not his fault, he said. It happened before the bank was involved. "It's like I bought some weed from a guy who bought it from another guy who bought it from another guy who bought it from a guy in Guatemala who killed someone for it," he said. But you knew, I argued back. Before he bought it, he knew where it had come from. He knew what his boon had cost someone else.

Climate change is often framed as a scientific or economic or environmental issue, not often enough as an issue of human justice. This, too, needs to change. From this moment on, many of us could get rich. Many of us could get high. Life will go on. Before it does, we should all make sure we understand the reality of what we're buying.

ACKNOWLEDGMENTS

She's too nice to think about it like this, but Jennifer Woo and I got to live only a few weeks of our new life together in Seattle before she was suddenly sharing me with a book project. I thank her for eventually agreeing to marry me anyway and for her near-infinite patience. This book, my first, took a long time, especially considering that it took her little more than nine months to produce a first who gives us much greater pride.

Without the early guidance of two people, I might not have understood the scale of the story I was chasing. My friend and first editor at *Harper's Magazine*, Luke Mitchell, sent me on my first trip to the Arctic—and then, once I was back, helped me see that the antics of gun-wielding Canadians were important not because of what they revealed about climate change or Canadian American relations but because of what they revealed about human nature. Heather Schroder, my agent at ICM, then helped me focus on the obvious but crucial fact that the warming, along with the humans it affects and who are affecting it, is global. The story of the book would have to be much bigger than the Arctic.

At Penguin, Eamon Dolan took a gamble on a first-time author and provided invaluable advice—certainly much more helpful than he could have known—before moving on to a new post. Virginia Smith replaced him as editor, steering a somewhat wayward project toward the end with a wonderful mix of humor and discipline. Her insights and willingness to dig deep on problems small and big made this a much better book. Kaitlyn Flynn, meanwhile, has done tremendous work to get us all the way to the finish line.

As I began reporting, the luckiest break I had was in hiring the journalist Damon Tabor to work with me as a researcher for a year. It didn't take either of us long to realize that I should have been the one working for him. Next time, I probably will be. Before he went on to bigger things, Damon made many of these chapters possible, whether by tracking down contacts or, in some cases, by identifying stories I didn't even realize were there. Also, he works out with bricks. Really.

Unwittingly and not, various organizations helped support my reporting or kept me and my family clothed as I completed this project. I owe a tremendous debt to Charles Eisendrath, Birgit Rieck, Mary Ellen Doty, Patty Meyers-Wilkens, Candice Liepa, and Melissa Riley of the Knight-Wallace Fellows at Michigan; to Oliver Payne, Peter Miller, Lynn Addison, Susan Welchman, Nick Mott, Marc Silver, Glenn Oeland, and Rebecca Martin of the National Geographic Society; to Esther Kaplan of the Investigative Fund at the Nation Institute; to Jon Sawyer and Tom Hundley of the Pulitzer Center on Crisis Reporting; and to Columbia University and the family of John B. Oakes. *Harper's Magazine,* which in addition to Luke Mitchell gave me excellent editors in the form of Genevieve Smith, Rafil Kroll-Zaidi, and Christopher Cox, funded and originally published versions of the chapters "Cold Rush" and "Too Big to Burn," along with a section of "Uphill to Money." One of my favorite magazine editors, Alex Heard of *Outside,* first commissioned what would become the chapter "Greenland Rising." A version of the chapter "Farmland Grab" was first published in *Rolling Stone,* where it was edited by the dogged and excellent Eric Bates.

For support of an often less tangible sort, I thank David and Duane Funk, Ronald and Lisa Woo, Grace Funk and Benson Wilder, James and Margaret Woo, Jason and Condor Woo, James and Nadine Harrang, and our friends in Seattle, New York, Eugene, Bellingham, and Ann Arbor.

Many hundreds of people spoke with me by phone, answered e-mail queries, or agreed to sit down for interviews. Some are named in the book. Most are not. I am deeply grateful to all. A few people went even further, allowing me to travel with them for days or weeks or to otherwise invade their lives to try to see the world from their perspective. Without the extraordinary generosity of Minik Kleist, Chief Sam, John Dickerson, Phil Heilberg, Pape Sarr, Enamul Hoque, Luke Alphey, and Nathan Myhrvold, I could not have written quite this book, and I would not have enjoyed the experience half as much. I hope I got it all right, and wherever I did not, any mistakes are my own.

Others I would like to single out for special thanks: Sergeant Strong, Dennis Conlon, Doug Martin, John Ferrell, Mead Treadwell, Peter Schwartz, Ron Macnab, Michael Byers, Scott Borgerson, Matt Power, Larry Mayer, Andy Armstrong, Brian Van Pay, Luciano Fonseca, Tasha Gentile, Jimmy Jones Olemaun, Alexander Sergeev, Artur Chilingarov, Luda Mekertycheva, Garrik Grikurov, Victor Poselov, Trine Dahl Jensen, Martin Jakobsson, Brenda Pierce, Dave Houseknecht, Jeremy Bentham, Adam Newton, Sverre Kojedal, Geoff Dabelko, Vanee Vines, Juliane Henningsen, Kuupik Kleist, Jens B. Frederiksen, Rikka Jensen Trolle, Nick Hall, Tim Daffern, Giora Proskurowski, Moshe Tessel, Rafi Stoffman, Abraham Ophir, Willi Krüger, Eric Gilliland, Marco Ernandes, John Winkworth, Joe Flynn, Paul Johnson, Susie Diver, Garry Wills, Bill Heffernan, Todd Shields, René Acuña, Stephanie Pincetl, Merlin Camozzi, Clay Landry, Bob Heward, Daniel Snaer Ragnarsson, Eric Sprott, Gudjon Engilbertsson, Jeremy Charlesworth, Jon Steinsson, Kenneth Krys, Kevin Bambrough, Ric Davidge, Serge Kaznady, Shirley Won, Sigrún Davíðsdóttir, Sverrir Palmarsson, Terry Spragg, Uli Kortsch, Sean Cole, Carl Atkin, David Raad, John Prendergast, Peer Voss, Phil Corzine, Phil Warnken, Jonathan Davis, Nate Schaffran, Nick Wadhams, Jenn Warren, Ethan Devine, Nkem Ononiwu, Abdoulaye Dia, Desneige Hallbert, Chad Cummins, Tim Krupnik, Clara Burgert, Noam Unger, Caroline Wadhams, Antonio Mazzitelli, Alessandra Giannini, Jean-Marc Sinnassamy, Gil Arias Fernandez, Simon Busuttil, Joseph Cassar, Darrel Pace, Wayne Hewitt, Josie Muscat, Ivan Consiglio,

Emmanuel Mallia, Atiq Rahman, Ryan Bradley, Rohit Saran, Ajai Sahni, Nazmul Islam, Atique Islam Chowdhury, Reza Karim Chowdhury, Binoy Bhattacharjee, Bibhu Prasad Routray, Samujjal Bhattacharjee, Jennifer Marlow, Jeni Krenciki Barcelos, Spencer Adler, D'lorah Hughes, Franco Maschietto, Piet Dircke, Peter Wijsman, Thijs Molenaar, Frans Barends, Richard Pelliccan, Rene Peusens, Jort Struik, Koen Olthuis, Conny van der Hijden, Marnix de Vriend, Daniel Pepitone, Johan Cardoen, Piotr Puzio, Susanne Benner, Paul Epstein, Rip Ballou, Thomas Scott, Danilo Carvalho, Michael Doyle, Mikki Coss, Chris Tittel, Emily Zielinski-Gutierrez, Alun Anderson, Greg Huang, Shelby Barnes, Marelaine Dykes, Casey Tegreene, Samuel Thernstrom, Lee Lane, Kenneth Green, David Schnare, Michael Ditmore, Aaron Donohoe, David Battisti, Neil Adger, Heather McGray, Roger Harrabin, Andy Hoffman, Andy Buchsbaum, and Richard Rood. Apologies in advance to those I've mistakenly left off this list.

I recruited friends and family members to help me brainstorm a title for this book, but publisher Ann Godoff bested all of us with *Windfall*. (To be fair to Ben Pauker, he came up with it, too, but I didn't notice.) Ben, Mike Benoist, Dave Shaw, Alex Heard, Japhet Koteen, Benson Wilder, Vanessa Gezari, Tim Marchman, Aisha Sultan, Mike Laris, Damon Tabor, Ethan Devine, Tamar Adler, Wilson Kello, Noam Unger, Kalee Thompson, James Vlahos, Adam Allington, Evan Halper, Madeleine Eiche, Kihan Kim, Aaron Huey, Giora Proskurowski: Thank you. I'm sorry your puns can't be published.

At the Seattle Public Library, Chris Higashi lent me a locker and a quiet room to write, for which I'm very grateful. Even closer to home, Aaron Huey and Kristin Moore moved in across the street, and Aaron— who was with me in Russia when I was readying the proposal for this book—offered me a spare desk in his office so he could see me finally complete it. I thank him as well as the good people of Empire, just down the block, who have provided yet another office away from the office, this one with coffee.

NOTE ON SOURCES

I reported this book in person over many years, and the majority of it is based on what I saw and heard. Wherever possible, I have checked the material against my thousands of pages of handwritten and typed notes and against the photographs and voice recordings I often make in the field. In many cases, I have been able to further confirm details by consulting others who were present.

Before and after my travels, I tried to read every article I could about each place and topic, and I owe much to the journalists who went before me and the news organizations—the *New York Times, Wall Street Journal, Financial Times, Washington Post, Los Angeles Times, Houston Chronicle, Christian Science Monitor*, NPR, BBC, *Guardian, Economist, Der Spiegel, Maclean's, Globe and Mail, Sydney Morning Herald*—still paying to send them out into the world. I have borrowed from their reporting and their ideas, and I keep digital copies of their articles in an overstuffed folder on my laptop's hard drive. Also invaluable were more localized sources of news. To name a few: *Barents Observer, Alaska Dispatch, Sermitsiaq, Haaretz, Imperial Valley Press, Africa Confidential, Le Soleil*, IRIN, ReliefWeb, *Times of Malta, Times of India, Daily Star*, and *Palm Beach Post*.

As I set out to understand the effects of climate change, I read *The Economics of Climate Change* by Sir Nicholas Stern (Cambridge, U.K.: Cambridge University Press, 2007), *Field Notes from a Catastrophe* by Elizabeth Kolbert (New York: Bloomsbury, 2006), *The Weather Makers* by Tim Flannery (New York: Atlantic Monthly Press, 2005), and *Six Degrees* by Mark Lynas (London: Fourth Estate, 2007). I later relied on

Animal Spirits: How Human Psychology Drives the Economy, and Why It Matters for Global Capitalism by George Akerlof and Robert Shiller (Princeton, N.J.: Princeton University Press, 2009) as I considered humankind's response to climate change.

The history of Arctic exploration and the Northwest Passage is covered in *Resolute* by Martin Sandler (New York: Sterling, 2006) and *Dangerous Passage* by Gerard Kenney (Toronto: Natural Heritage, 2006). To understand Canada's uneven relationship with its own north, I read *Canada's Colonies* by Kenneth Coates (Toronto: Lorimer, 1985) and *Tammarniit (Mistakes)* by Frank Tester and Peter Kulchyski (Vancouver: University of British Columbia Press, 1994).

For a month in the frozen Chukchi Sea, scientists and State Department representatives aboard the U.S. icebreaker *Healy* kept me and themselves entertained with informal lecture nights—the source of much of what I have learned about the Law of the Sea, the melting polar ice cap, and the jockeying between various coastal states for control of the Arctic and its oil-rich seabed. Chief Scientist Larry Mayer, the director of the Center for Coastal and Ocean Mapping at the University of New Hampshire, has been a particular resource. In Russia, Yuri Kazmin provided further insights, as did Canada's Ron Macnab, Denmark's Trine Dahl-Jensen, Sweden's Martin Jakobsson, and other sources in Washington and Moscow who would prefer not to be identified. At the U.S. Geological Survey, Don Gautier, Brenda Pierce, and Dave Houseknecht helped me fathom the potential size of the petroleum prize.

The history of Shell's scenario planning is covered in *The Art of the Long View* by Peter Schwartz (New York: Doubleday/Currency, 1991) and a follow-up, *Learnings from the Long View* (Seattle: CreateSpace, 2011). Also helpful are Shell's many public reports and the various writings of Art Kleiner, author of *The Age of Heretics* (New York: Doubleday/Currency, 1996).

To learn about California's wildfires and their context, I read *The*

Control of Nature by John McPhee (New York: Farrar, Straus and Giroux, 1989), *The Phoenix* by Leo Hollis (London: Phoenix, 2009), and *A Discourse of Trade* by Nicholas Barbon (London, 1690). California and the American West's never-ending struggles against drought are documented in *Cadillac Desert* by Marc Reisner (New York: Viking, 1986), *Unquenchable* by Robert Glennon (Washington, D.C.: Island Press, 2009), and *California: A History* (New York: Modern Library, 2005), the historian Kevin Starr's distillation of his seven-part series on the Golden State and the American dream.

Along with current and former U.S. government sources, *Emma's War* by Deborah Scroggins (New York: Pantheon, 2002), *The Root Causes of Sudan's Civil Wars* by Douglas H. Johnson (Bloomington: Indiana University Press, 2003), and *Atlas Shrugged* by Ayn Rand (New York: Random House, 1957) were my guides to Phil Heilberg's patch of Africa. For an overview of global food crises, I turned to *The Coming Famine* by Julian Cribb (Berkeley: University of California Press, 2010) and *An Essay on the Principle of Population* by Thomas Malthus (London: J. Johnson, 1798). To understand the history of shelterbelts like the Great Green Wall, I read *Woman Against the Desert* by Wendy Campbell-Purdie (London: Victor Gollancz, 1967).

The amphibious future envisioned by Koen Olthuis is detailed in his book *Float!* (Amsterdam: Frame, 2010), written with David Keuning. The rise of infectious diseases in a warmer world is described in *Changing Planet, Changing Health* by the late Paul Epstein and Dan Ferber (Berkeley: University of California Press, 2011).

I was an early (and always silent) member of a lively Google Group discussing geoengineering that was started by Ken Caldeira, which gave me insight into the characters and motivations that would birth two excellent books as I was wrapping up my own: *How to Cool the Planet* by Jeff Goodell (Boston: Houghton Mifflin Harcourt, 2010) and *Hack the Planet* by Eli Kintisch (Hoboken, N.J.: Wiley, 2010). *SuperFreakonomics* by Steven Levitt and Stephen Dubner (New York: William Morrow,

2009) helped explain the inner workings of Intellectual Ventures, while *Fixing the Sky* by James Rodger Fleming (New York: Columbia University Press, 2010) offered anecdotes and a much-needed reminder that we have always wanted to control the weather.

Geoengineering lectures I attended in 2010 and 2011 at the University of Washington attracted some of the nascent field's best scientific and ethical minds: Fleming, David Keith, Dale Jamieson, Phil Rasch, Alan Robock, Jane Long, Christopher Preston, Steve Rayner, Ben Hale, and Michael Robinson-Dorn. Often in attendance was University of Washington professor David Battisti, who gladly discussed both the science and the intrigue of geoengineering with me. Stephen Gardiner, a philosophy professor who organized the lecture series, is also the author of *A Perfect Moral Storm: The Ethical Tragedy of Climate Change* (New York: Oxford University Press, 2011). His writings helped me understand that contrary to conventional wisdom, global warming is not a classic "tragedy of the commons" as first described by the ecologist Garrett Hardin—or at least that if it is, some of the metaphorical herdsmen among us have bigger cows.

When I traveled a second time to Alaska's Chukchi Sea and stayed in the village of Point Hope, I carried with me *The Firecracker Boys* by Dan O'Neill (New York: St. Martin's Press, 1994), the story of how we nearly detonated six hydrogen bombs to create a new Arctic harbor—a brilliant history I wish I had read long ago.

Lastly, a note on translations: Some in the book are my own. For dialogues originally in Russian or French, I have done my best to capture the speakers' meaning—but rarely can I capture their eloquence. The few phrases originally in Spanish are better rendered.

INDEX